Bionanotechnology

Johnathan L. Kiel

Bionanotechnology

Radio Controlled Antimicrobial and Genetic Vectors

Springer

Johnathan L. Kiel
Bioanalysis Consulting LLC
Universal City, TX, USA

ISBN 978-3-031-63005-7 ISBN 978-3-031-63006-4 (eBook)
https://doi.org/10.1007/978-3-031-63006-4

This Springer imprint is published by the registered company Springer Nature Switzerland AG
The registered company address is: Gewerbestrasse 11, 6330 Cham, Switzerland

Dedicated to the millions infected, and those who have died from COVID, and the military scientists and engineers who contributed to the development of the science and technology herein that may avert such tragedy in the future, and, finally, to all those scientists, veterinarians, physicians, nurses, and other health professionals who are in the Global One Health battle against emerging and re-emerging infectious disease, and antimicrobial resistance.

Preface

The scientific foundations of this work are described in the CRC Press, Taylor & Francis Group, book *Type B Cytochromes: Sensors and Switches* (1995 and, lastly, as a Revival CRC Press edition in 2018). The popular science version of the history and why this material was developed are described in detail in the 2012–2018 collection, The Black Dragon Trilogy (Amazon eBooks, 2018: *Notes from the Shadows* (2012, 2015), *Pathogenic Ecology* (2015), and *Nanowarfare* (2016)). Where these books tell how it was made, this book describes how to make it (DALM) and the scientific basis for the properties and their practical applications in antimicrobial activity and gene delivery and modification.

Finally, this book puts the pieces together into how this material defies conventional classic electromagnetic behavior and how the scientific basis for it can be generalized beyond this single material.

The specifics in computation and theory will be addressed by reference to more specialized texts and references. The aim here is not to duplicate such work but emphasize how it connects to interdisciplinary studies that led to the resolution of the paradox discussed herein. All representations of equations, figures, and pictures are either original or have been produced by a group consisting of both US government employees and non-US government employees, and as such is subject to 17 U.S.C. Sec. 105 and exempt from copyright and are in the public domain. All the information has been previously approved for unlimited public release by USAF Public Affairs and any potential copyright or proprietary conflicts resolved by permission.

Universal City, TX, USA Johnathan L. Kiel

Acknowledgments

Thanks to the USAF Air Force Research Laboratory Counterproliferation Team, especially Dr Eric Holwitt, Dr Jill Parker, Dr John Bruno, Dr Mark Sloan, Dr Jeevalatha Vivekananda, Dr Maomian Fan, Mr Gerald O'Brien, Mr John Alls, Mr Richard Weber, Mr Homer Gifford, Ms Lucy Stribling, Mr Pedro Morales, Ms Veronica Sorola, and many more collaborators and contractors too numerous to list.

Contents

List of Figures

List of Tables

List of Equations

About the Author

Johnathan L. Kiel received his BS and DVM, Summa Cum Laude, 1974; has been a Diplomate of the American College of Veterinary Microbiologists, since 1984; and was elected a Charter Diplomate in the specialty of Veterinary Parasitology in 2011. He is also a Fellow of the American Association for the Advancement of Science, elected 22 September 1995, and a Fellow of the Air Force Research Laboratory elected 21 October 1998. He has authored 106 publications including peer-reviewed journal articles, proceedings, government reports, book chapters, and four books before this one. He is an inventor and holder of 30 US patents. Dr Kiel served 35 years of combined active-duty military and civilian service in the USAF. He served in the active-duty military 13 years, starting as a Captain in the Veterinary Corps (remaining in the Corps until it was disbanded April 1980), serving as a Base Veterinarian from 1975 until 1977, caring for 25 military working dogs, military member and dependent owned animals, and USAF stabled horses. He completed his military service as a Major, Regular USAF, Biomedical Sciences Corps. He then served in civil service (22 years) in the Air Force, until May 2011. Before that he was a 2nd Lt, from January 1973 till April 1975, in USAF Reserves, Medical Services Corps, Texas A&M, College of Veterinary Medicine. He taught pathogenic bacteriology at Texas A&M University School of Veterinary Medicine from 1974 until April 1975 when he entered military active duty in the USAF Veterinary Corps. He received his PhD in Microbiology and Biochemistry from Texas Tech Health Sciences Center, School of Medicine, in 1981. He was the AFRL Senior Scientist in Electromagnetic Bioeffects and Chem/Bio Counterproliferation from 2002 to 2011. He retired on May 2011 at the level of Senior Scientist in the Senior Executive Service, a civilian equivalent of a Brigadier General.During his tenure at the Air Force Research Laboratory and School of Aerospace Medicine, he taught veterinarians, physicians, nurses, and medical and entomological technicians about snake venom toxicology, field and medical treatments of the associated injuries, and infectious disease such as anthrax as part of Force Health Protection continuing education (approved as veterinary, medical, and nursing CE for licensure). During his time with the USAF, he also consulted on and treated many snake and other reptile clinical cases and wrote articles and book chapters on their care and diseases. He also led a research team in basic research on non-ionizing radiation bioeffects, and in basic and applied research in biosurveillance and counterproliferation of Select Biological Agents (Especially Dangerous Pathogens) and other high-impact zoonotic diseases. His team fielded an identification and viability detection kit for all the Services for anthrax (and provided the initial field training), just in time for the first Iraqi War and recovered the last *Bacillus anthracis* out of Iraq before the second war in March of 2003. His team provided essential feedback to the CDC and Department of Defense on their Select Agent regulations, as far as their maintenance and oversight of transfer. He was a Responsible Official who oversaw the commissioning and operation of a BSL-2+ and then a BSL-3+ (with capability to collect, isolate, and safely transfer level 4 agents as well as maintain and work with an extensive archival collection of BSL-3 Select Agents and other especially dangerous pathogens) microbiological laboratory, both registered under the Select Agent Program from 5 January 2001 until July 2010. He implemented their closure due to the Base Realignment and Closure Act in 2010, helped design the new USAF BSL-3 facility at Wright-Patterson AFB, Ohio, and secured transfer of the biological agents to that laboratory. His team performed experiments involving plague, tularemia, Q fever, spotted fever, brucella, vaccinia, retroviruses, Venezuelan Equine Encephalitis, and subunit replicons of Ebola virus, and anthrax bacteria, and toxins and select and non-select emerging pathogenic agents with a perfect safety and security record from 1989 until the lab's closure. His international collaborations extended from the United Kingdom, Australia, South Africa, Vietnam, Argentina, and France to the French West Indies (World Organisation for Animal Health (OIE) Heartwater Laboratory) and as a consultant to Southwest Research Institute as part of the Department of Defense, Defense Threat Reduction Agency's Cooperative Biological Engagement Program in Azerbaijan from 2011 to 2015 and Kazakhstan from 2015 until 2016.

Abbreviations

θ_ε	Einstein's temperature
3AT	3-amino-L-tyrosine
APC	antigen-presenting cells
ASExp	aptamer selection express
ASK1	apoptosis signal regulating kinase
AT	adenine/thymine
ATP	adenosine triphosphate
BAS	*Bacillus anthracis* Sterne
Bt	*Bacillus thuringiensis*
CNT	carbon nanotube
CRISPR	clustered regularly interspaced short palindromic repeats
C_v	heat capacity at a fixed volume
DALM	diazoluminomelanin
DC	direct current
DNA	deoxyribonucleic acid
E	electric field
E. coli	*Escherichia coli*
ECT	electroconvulsive therapy
EGFR	epidermal growth factor receptor
ELASA	enzyme-linked aptamer sorbent assay
ELISA	enzyme-linked immunosorbent assay
ESR	electron spin resonance
FAD	flavin adenine dinucleotide
Ft	*Francisella tularensis*
GC	guanine/cytosine
GFP	green fluorescent protein
GHz	gigahertz (10^9 Hertz)
IKK	inhibitor of nuclear factor kappaB (I*k*B)
JNK	C-Jun N-terminal kinase
KE_{avg}	average kinetic energy
KE_{mp}	most probable kinetic energy

kV	kilovolt (1000 volts)
LPS	lipopolysaccharide
MAPK	mitogen-activated protein kinase (K = kinase)
NAD(H)	nicotinamide dinucleotide (hydride)
NADPH	nicotinamide adenine dinucleotide phosphate hydride
NEMR	nonionizing electromagnetic radiation
NMDA	N-methyl-D-aspartic acid
NO	nitric oxide
NOS	nitric oxide synthase
NP	nanoparticles
PBS	physiological phosphate-buffered saline
PCR	polymerase chain reaction
RFR	radiofrequency radiation
RNA	ribonucleic acid
RNAi	Ribonucleic acid interference
rTMS	transcranial magnetic stimulation
SAR	specific absorption rate of energy
SELEX	systematic evolution of ligands by exponential enrichment
SEM	scanning electron microscopy
SNA	small synthetic nucleic acids
SQUID	superconducting quantum interference device
SWCNT	single-walled carbon nanotubes
TADF	thermally activated delayed fluorescence
Tdm	transition dipole moment
TEM	transmission electron microscopy
TLR-9	toll-like receptor 9
TNF	tumor necrosis factor
TSB	tryptic soy broth
σ	conductivity

Chapter 1
Introduction

Abstract The greatest challenge in science is a paradox. The greatest satisfaction is in resolving it. A perusal of the contents of this book may mislead the reader to think that it is about a specific material for biological manipulation using electromagnetic radiation, specifically radiofrequency and microwave radiation. Although that view contains some truth, the intent here is to be much more expansive.

Keywords Paradox · Thermodynamics · Quantum · Phenomenology · Electromagnetics · Thermochemiluminescent · Radiofrequency · Microwaves · Nanomaterials · Dosimetry

This book is about the interdisciplinary, multidisciplinary approach to solving a much larger problem, a paradox of scientific theory and observation. The paradox is classical electromagnetics and thermodynamics versus quantum theory (mechanics). Is radiofrequency to microwave radiation acting on diazoluminomelanin reactions as time-varying fields, through secondary heating, or as quanta being specifically absorbed?

The application of the resolution is to greatly improve the remote manipulation of nanomaterials as vectors of genetic modification and antimicrobial effects. This resolution of the paradox would not be possible with a single discipline. However, a multidisciplinary approach is fraught with communication, comprehension, and cultural discontinuities and disparities that must be resolved by some common scientific language and, at last, a limited mutual understanding by those of each other's disciplines. The bridging of these gaps in the specific examples given is one goal of this book.

No one can be an expert in all things, but a common language among the disciplines is necessary to solve a paradox, and, in particular, the one at hand, and needs to be established. The glasses or filters through which each sees this world need to be merged into one to resolve the paradox. There is a discipline that attempts to dispassionately examine this melding process. It is called phenomenology.

The term has two meanings. The one in physics is all that can be measured through experimentation. The second, founded by Edmund Husserl (April 8,

J. L. Kiel, *Bionanotechnology*, https://doi.org/10.1007/978-3-031-63006-4_1

1859–April 27, 1938), is a philosophical approach examining how we interpret observations and measurements as well as experience and consciousness. He founded this philosophy in the twentieth century at the Universities of Gottingen and Munich, Germany. For our purposes, it is the collective filters through which measurements were made by the individual disciplines that led to the paradox to be examined and how the filter must be changed to resolve it.

A relevant example with very broad application is how to resolve classical relativity with quantum field theory to yield quantum gravity. The hang-up here is the comparison of the phenomena of very large structures, like stars and galaxies, dominated by gravity with the quantum mechanical fundamental phenomena of very small structures like atoms, molecules, and even subatomic particles. These appear to be negligibly affected by gravity until the density of the gigantic structures collapses under their own gravity then they become like very dense small structures. Perhaps what is missed is not the differences due to mass but the similarities of their respective derivative properties of highly dense forms. When the mass of a star collapses into an ultra-dense black hole, the chemical and physical properties, determined by quantum properties, disappear, except for mass and perhaps spin. Even photons cannot escape the black hole.

Phenomenologically, the classical and quantal characteristics have merged into something else. The result is barely describable but connected by extreme density. The observation of a black hole is the measurable but incomplete observable (at least through our disciplinary biases) resolution of the paradox of classical relativity and quantum theory. Energy and matter are converged into something else.

The argument from analogy can be very dangerous in drawing conclusions about other processes because superficial similarity does not always translate into causality. However, in this case, translating from the large to the small is also a problem that yields the paradox of consideration of this book. Classical electromagnetics involves circuits, antennas, and waves, but those macroobjects are still composed of micro-, nano-, molecular as well as atomic and subatomic components that play a role in electromagnetic responses. Oddly enough, like in the case of black holes, in classical relativity and quantum theory, in the disappearance of chemical and physical properties, energy density plays a critical role in our problem at hand.

To understand how we arrived at the paradox to be discussed, one must consider the historically evolving needs and requirements of the military in electromagnetic bioeffects and their extension to countering biological and chemical warfare agents. The evolving needs and requirements can be summarized as follows.

The first military goal was to create a radiofrequency (microwave) radiation sensor that required no intrusive instrumentation (instead using visual or wireless remote optical sensing) to reveal radio-frequency radiation (RFR) energy absorption and to what degree. Luminescence or fluorescence was the most opportune reporting modality because it could be quantitatively observed and be remotely recorded and measured with available instrumentation. We developed instruments to quantify low-light levels.

Next, the reporter needed to be nonbiological but able to be incorporated into tissue equivalent material (saline, at a minimum, to more complex gel compositions for semi-solid tissue equivalent "phantom" material) that would allow the spatial and temporal measurement of responses. Of course, the material and the animal or human models (phantoms) that held it must be radiotransparent as well as optically transparent. This was required so as not to disrupt the RFR and optical signals, respectively. The material had to hold a constant signal for a constant input of energy and/or indicate a change in temperature from such input.

It followed that the reaction generating the signal, indicating the amount of RFR absorption, needed to be simple in composition and execution for nonchemists. The material should minimally perturb the dielectric properties of the tissue-equivalent material. It needed to be of low toxicity, be easily manufactured and stored, and have a reasonable shelf life and reproducible performance. Eliminating activating and catalytic components was essential for acceptance at large.

Then with the emergence of biological warfare concerns, the adaptation of this indicator as a reporter for a specific sensor for their presence became necessary. This goal was accomplished by using the durable luminescence and fluorescence of the polyphenylenes arising from the RFR dosimetry research.

The need to eliminate organic solvents and increase the yield of this luminescent polymer, in the shortest possible time, led to the discovery and development of bacteria that could biosynthesize the polymer in a fermenter and then the isolation of genes for the metabolism that generated it. These were cloned into bacteria to make them super producers of the polymer.

The requirement for a viability assay and selective medium for the number one biological warfare threat, anthrax, led to adapting the biosynthetic method to the rapid and selective viability, self-labeling, and identification methods for use in the battlefield, under less-than-ideal isolation and cultivation conditions.

The isolation of genes for the biosynthesis of the RFR-sensitive polymer, without killing the host, allowed for the biosynthesis of the polymer in situ in animal and human cells. On the RFR side, expanding the investigations using the activated polymer to high power, short pulse RFR and microwaves, ultrasound, and direct current fields revealed characteristics beyond just thermal responses of the luminescence.

The sensory properties were extended to killing bacterial pathogens by pulsed microwaves. This action was accomplished at energies (low duty factors and exposure times) that would minimize energy requirements and collateral damage.

The final requirement was to extend these effects to inactivating viral pathogens and to eliminate the requirement for activating reactants with what was endogenously contained in the material. The antiviral need required the material to be made in situ in cells infected with the virus or delivered in a convenient dispensable form on a nanoscale. That form needed to be remotely activated with electromagnetic radiation. This requirement was incompletely accomplished because of the ending of the military research program. This book addresses what needs to be done to complete this last goal and expand the applications to meet the needs of the general population.

Chapter 2
Foundation Principles and Dosimetry of Radiofrequency Radiation

Abstract This chapter discusses the foundational work on the appropriate measurements of nonionizing electromagnetic radiation (NEMR) absorption and interaction that are relevant to human health, safety, and bioeffects, whether innocuous or pathogenic. The United States Air Force (USAF) and collaborating academic researchers (especially from the University of Utah, Department of Electrical Engineering) made a major contribution (perhaps the most important in the field) to understanding and implementing the relative measurements of NEMR for health and safety purposes in the form of several editions of the Radiofrequency Radiation Dosimetry Handbook. The fourth edition is an elegant teaching text for those who are not electrical engineers or physicists. It is a necessity for biological and biomedical researchers interested in the health and physiological effects of NEMR and who wish to make measurements of doses of NEMR that are relevant to the real world and well defined in terms of exposure dosimetry. In this chapter, I summarize this work and show its relevancy to the problem at hand.

Keywords Handbook · Electric · Vector · Magnetic · Electromagnetic · Maxwell's equations · Coulomb's law · Phantom · Dosimetry · Quantum · Duality · DeBye equation · Standards

Without good dosimetry and animal-to-human and vice versa extrapolation of exposures, the biological data of effects would be nearly useless. The newest edition of the Radiofrequency Dosimetry Handbook (5th Edition, 2009), published by the United States Air Force (USAF), refined the measurement methodologies with more modern instrumentation and methods [1]. However, the fourth edition is by far the best presentation of basic electromagnetic radiation physics and engineering [2]. This chapter is intended not to reproduce these elegant texts, their instruction, and the descriptions of electromagnetic radiation theory and dosimetry but to draw on the information by reference when it is most appropriate in addressing the intersections of macroscopic vs micro- and nanoscale mechanisms.

At least an introductory presentation of the mathematics and theory used by physicists and electrical engineers, as the universal language, to describe

J. L. Kiel, *Bionanotechnology*, https://doi.org/10.1007/978-3-031-63006-4_2

electromagnetic effect mechanisms is necessary at this point. They define the classical mechanism that contributes to the paradox. Where classical electromagnetics attributes bioeffects to the thermalization of absorbed electromagnetic frequencies and electric shock at very low frequencies, the observation of diazoluminomelanin bioeffects increasing exponentially when the exposure is cooled defies these classical explanations. This contradiction merits an examination of the potential quantum effects in radiofrequency radiation (RFR) interactions.

The handbook illustrates the great utility of a classical explanation that is highly appropriate and useful in predicting, within set boundaries, the relevant dosimetric measurements for health and safety consequences and permissible standards of exposure in almost all cases. However, the texts lack a fundamental connection, except by analogy, to the molecular and atomic effects of nonionizing electromagnetic radiation (NEMR). Classical electromagnetic theory, calculations, and the measures discussed apply wave theory and charge interactions to explain molecular effects.

In contrast, in the quantum world, photons are discrete energy packets and waveforms that interact with matter in a very special way, unlike black bodies, which interact with thermal intensity in a continuous spectral way. The duality of quanta and discontinuity of their interactions with matter is found in the mathematical approach that dominates the discussion of scalar vs vector quantities, in quantum and chemical terms, particle (discrete photon energies) vs dynamic (kinetic energy and waveforms). A simple example of the interchangeable nature of scalar vs vector is in temperature compared to kinetic energy. Temperature at a given point (spatial and temporal) is considered scalar, but on the molecular level, it is equivalent to the average kinetic energy of molecular or atomic particles, vectorial, (because in the equation, velocity is a vector quantity) (Eq. 2.1).

Equation 2.1: A and B Kinetic Energy

$$\mathbf{KE_{mp}} = \mathbf{\tfrac{1}{2}(RT)}$$

$$\mathbf{KE_{ave}} = \mathbf{(3/2)(RT)}$$

The equations of most probable KE_{mp} and average kinetic energy KE_{ave}, with differing degrees of freedom (1 vs 3).

The factors of the relation are both scalar and vectorial, mass and velocity, respectively. This example is also our first example of philosophical phenomenology, where one's frame of reference, or worldview, determines whether the data reveal a classical or quantal effect. As long as the measurement works to support classical actions and predicted responses, the effect will always be interpreted as classical. Both can exist simultaneously if the measurement, at the resolution necessary for a particular mechanism, would only reveal a new imbedded one if the resolution of the measurement were changed sufficiently. Even with classical electrodynamic and thermodynamic behaviors (averages of molecular and electronic behaviors), the measurement of the effect of an electric or magnetic field applied is influenced throughout the object affected by surrounding induced electromagnetic fields.

As in quantum mechanical measurement, the measurement influences the state of a molecule or atom in a quantum property, and they, in turn, influence each other. These self-correlating influences and the cross-influence of classic electromagnetic field and charge theories and quantum behavior are the very intersections that this text intends to analyze. They will be analyzed and reconciled to explain the chemical and physical behaviors of certain molecules and nanostructures in response to applied NEMR.

Remarkably, the quantum mechanical designations of the electronic positions (energy level) in atoms and molecules contain kinetic and potential energy values that correspond to the probability of finding an electron in a location. These values are discrete, but the locations and momentums of the electrons are probabilities. Neither the momentum (mass times velocity) nor the location (potential energy) can be determined simultaneously.

The wave functions of a particular electron or electrons are not a wave but a probability function with discrete eigenvalues (nonzero) describing the electron's probable position and energy, characteristics that determine its loss or gain of a photon, of a particular energy, to displace it.

The positive (as an electron absent hole) and negative charges (an electron or a pair of electrons) in classic theory are subject to continuous field displacement from neutrality in an atom or molecule and are not found by the need for discrete energies for the displacement. This is because they represent either widely shared electrons in the conduction band (orbitals) of metals, other conductors or semiconductors, or whole atomic or molecular charged units. I will briefly discuss classic electromagnetics here and leave the details in the handbook.

This approach fails to describe the anomalous behavior in our paradox. However, it still plays an important role in the counterintuitive chemical and physical behaviors demonstrated in the specific phenomenology (the physics interpretation when philosophical phenomenological biases are removed).

Classical macroscopic electrodynamics are what are imposed on the quantal effects observed in some basic explanation, with details given by reference to the ones in the handbook. We will start with fundamentals that apply to all electromagnetic radiation and that represent the first formal combination of forces in physics. The essence of this principle is found in Maxwell's equations. Unfortunately, they are very difficult to solve, so physicists and engineers have resorted to parallel approximating methods, with certain boundary conditions, to provide workable solutions for engineering applications.

Maxwell's equations [3] are shown in the equations in Eq. 2.2. In these equations, the direction of forces and displacement (and their interactions on a charged particle) are of the utmost importance. Forces may be static and act independently on particles or be coupled to displacement (velocity) or changes in displacement and/or velocity (acceleration). Maxwell's equations (differential, Eq. 2.2 A, and integral, Eq. 2.2 B, forms) link magnetic and electric fields together when they are in motion to form a single force, the electromagnetic force.

Equation 2.2: A and B Maxwells Equations

A. **The partial differential form:**

$$\nabla \cdot \vec{D} = \rho,$$
$$\nabla \cdot \vec{B} = 0,$$
$$\nabla \times \vec{H} = \vec{J} + \frac{\partial \vec{D}}{\partial t},$$
$$\nabla \times \vec{E} = -\frac{\partial \vec{B}}{\partial t}.$$

$$\nabla = \vec{e}_x \frac{\partial}{\partial x} + \vec{e}_y \frac{\partial}{\partial y} + \vec{e}_z \frac{\partial}{\partial z} = \left(\frac{\partial}{\partial x}, \frac{\partial}{\partial y}, \frac{\partial}{\partial z}\right)$$

Where the coordinates are x, y, and z and the standard basis or unit vectors of axes are

$$\vec{e}_x, \vec{e}_y, \vec{e}_z.$$

A standard basis vector is a one-unit long vector that points directly along or in parallel to the corresponding axis.

B. **The integral form in a closed loop applied to Gauss's law of electricity, Gauss's law of magnetism, Faraday's law of induction, and Ampere's law plus Maxwell's displacement current, respectively:**

$$\oint \vec{E} \cdot d\vec{A} = \frac{q}{\varepsilon_0}$$
$$\oint \vec{B} \cdot d\vec{A} = 0$$

$$\oint \vec{E} \cdot \vec{ds} = \frac{d\Phi_B}{dt}$$

$$\oint \vec{B} \cdot d\vec{\ell} = \mu_0 \left(I + \frac{d}{dt}\left(\varepsilon_0 \int \vec{E} \cdot d\vec{A}\right)\right)$$

$$\oint \vec{H} \cdot d\vec{l} = I_{\text{enc}}$$

Gauss's law of electricity is that the integral of the outgoing electric field over an area enclosing a volume equals the total charge inside.

Gauss's law of magnetism is that no magnetic charge exists, therefore no "monopoles."

Faraday's law of induction is that which gives the total voltage change around a circuit, which is generated by a varying magnetic field going through a circuit.

Ampere's law (last curl integral) states that the magnetic field generated by an enclosed electric current (I_{enc}) is proportional to the size of that electric current with a constant of proportionality equal to the permeability in free space.

The definitions of the symbols are as follows:

dA = an infinitesimal part of an area
ds = an infinitesimal part of a surface
dl = an infinitesimal part of differential length all above in a Cartesian system
I = current
B = magnetic flux density
H = fixed magnetic field intensity
Φ_B = magnetic flux
E = electric field
D = displacement
μ_o = permeability
J = free-current density, A/m^2
P = free-charge density, C/m^2
C = Coulombs
A = amperes; 1 ampere = $\mu_o/2\pi$
m = meter

A major characteristic of Maxwell's equations is that they show that time-varying magnetic fields (field flux B) are equivalent to time-varying electric (E) fields (ion or electron displacement); in other words, a conductor mechanically moving through a fixed magnetic field (H) induces a current in the conductor (wire), as in a generator. H is the magnetic field that is integrated around a closed path in which a current is enclosed by that path (Ampere's law). Conversely, a direct current (DC) passing through a coil around a ferromagnetic core induces a magnetic field. The process of E-field to H-field conversion is used in a step-up (voltage) or step-down (voltage) transformer (with alternating current).

In Maxwell's equations, there is a dot product of vectors for a dot $b = c$. Equation 2.3 mathematically defines a dot product, and Fig. 2.1 shows its graphical representation. The cross product is mathematically defined in Eq. 2.4 with the graphical representation in Fig. 2.2.

Del dot-E product equals dispersion, and del cross-E product equals curl. Dispersion occurs when sinusoidal waves (electromagnetic waves) of different wavelengths have different propagation velocities, leading to wave packets of mixed wavelengths that spread out over space. Curl is the operation that when applied to a vector field measures the circulation of that field in a confining space or volume (bounded).

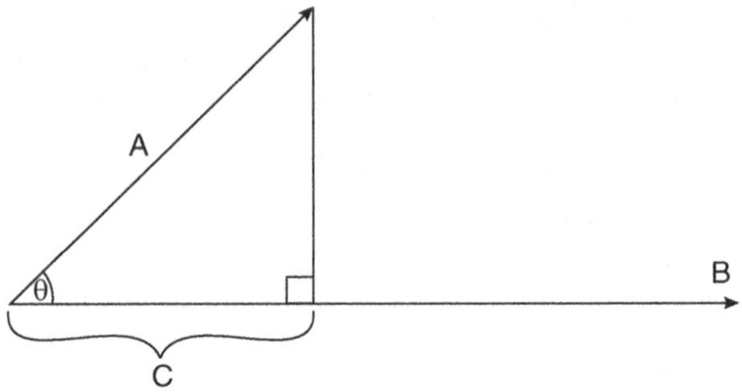

Fig. 2.1 Graphic representation of a vector dot product

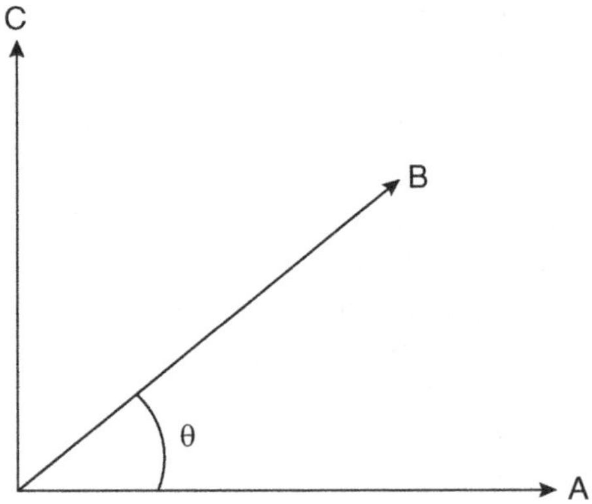

Fig. 2.2 Graphic representation of a vector cross product

The physical meaning of the dot and cross products is that the dot product gives the work done (force causing or changing the displacement of an object), whereas the cross product gives torque (force like turning a screw). Mathematically, one can think of their multiplying the magnitudes of two vectors, the dot product being a scalar quantity, indicating magnitude, not direction, while the cross product is a vector, which indicates magnitude and direction. Another way of looking at them is that a dot product measures how much two vectors point in the same direction and the cross product measures how much two vectors point in different directions.

Maxwell's equations are not just a mathematical explanation of how magnetic fields are converted into electric fields but also the first step toward how electromagnetic fields are converted into work (as in an electric motor) or work into electricity (as in a generator). As the first law of thermodynamics states, work is energy and equivalent to heat (as in a microwave oven or the classical electromagnetic interaction with molecules).

Equation 2.3: Dot Product

$$\mathbf{a} \bullet \mathbf{b} = \|\mathbf{a}\| \; \|\mathbf{b}\| \; \cos\theta$$

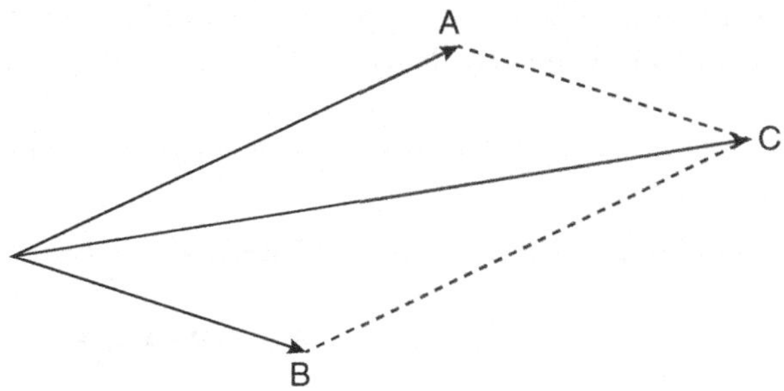

Fig. 2.3 Graphic representation of vector addition

Equation 2.4: Cross Product
For the vector cross product $A \times B = C$:

$$A \times B = \|A\| \; \|B\| \sin\theta n$$
$$\|A\| = \text{length of Vector A}$$
$$\|B\| = \text{length of Vector B}$$
$$\theta = \text{angle between A and B}$$
$$n = \text{unit vector perpendicular to the plane containg a and b}$$

The vector addition is given by the Pythagorean theorem, expressed in Eq. 2.5 and graphically represented in Fig. 2.3.

Equation 2.5 Vector Addition: Pythagorean Theorem

$$c = \sqrt{a^2 + b^2}$$

All electromagnetics are based on the forces (F) that electric charges (q or Q) impose on each other (Eq. 2.6), Coulomb's law (k_e is the Coulomb constant ≈ 8.988×10^9 N·m^2·C^{-2}). When these forces are moved, they yield work and energy.

Equation 2.6: Coulomb's Law

$$|F| = k_e \frac{|q_1 \, || \, q_2|}{r^2}$$

The concepts of permittivity (ε) and permeability (μ) will be discussed here at the macroscale and delved into at the molecular and atomic levels in the next chapter. In the former context, q = one charge force resulting from the presence of another charge, q_1 vs q_2, or Q. R (carrot) is a unit vector pointing along a line from Q to q. R is the distance between the two charges. The product of the charges and the distance between them is the dipole moment. Permittivity is the property of a material that measures the opposition against an electric field. The dielectric constant is the ratio of the permittivity of the substance to the permittivity of the free space, $\varepsilon_{0,}$ (farads/m). Each charge can have the same sign (repulsion) or different charges (attraction). There can be many of these charges with the bulk effect being the *E* field (field strength vector), the total force generated by all the charges affecting each other based on the distance between the charges (Eqs. 2.7 and 2.8). A very small charge placed in this field of charges generates a force (*F*) on this small test charge.

Equation 2.7: Electric Field of Interacting Charges

$$\mathbf{E} = \mathbf{Q}\hat{\mathbf{R}} / 4\pi\varepsilon_0\mathbf{R}^2$$

Equation 2.8: Force Generated by Charges

$$\mathbf{F} = \mathbf{q}\,\mathbf{E}$$

As an electric charge moves, a force is generated perpendicular to the range of motion, the magnetic flux density (*B* field).

v is the velocity of q, and F_m is the force perpendicular to this charged particle velocity. The F_m direction is described by Eq. 2.9.

Equation 2.9: Magnetic Field of Moving Charges

$$\mathbf{B} = \mathbf{F}_{\mathrm{m}} / \mathbf{qv}$$

This is the cross product perpendicular to both *v* and *B*. According to the right-hand rule, it is in the direction of a screw being turned into *B*. The total force in any product in the *E* and *B* fields together is expressed in Eq. 2.10.

Equation 2.10: Force of Moving Charges in an Electric Field

$$\mathbf{F} = \mathbf{q}\left(\mathbf{v} \times \mathbf{B}\right)$$

If these fields are not moving, that is, static, then the *E* and *B* fields are independent of each other and do not generate new *E* and *B* fields when an object interacts with them (scattered fields) that would add to the incident *E* field (Eq. 2.11).

Equation 2.11: Total Force of Moving Charges in Electric and Magnetic Fields

$$\mathbf{F} = \mathbf{q}\left(\mathbf{B} + \mathbf{v} \times \mathbf{B}\right)$$

The energy absorbed (a dose) is connected to internal fields and is strongly related to the dimensions of the object, as in Eq. 2.12.

Equation 2.12: Internal Electric Field of Absorbed Electromagnetic Energy

$$\mathbf{E}_{\text{in}} = \sqrt{\frac{\rho_{\text{m}}\mathbf{SAR}}{\omega\varepsilon_0\varepsilon''}}$$

where P_{m} equals the mass density in kg/cubic meter.

The solutions of Maxwell's equations by frequency dependency and matched systemized techniques are based on the wavelength of the NEMR. In selecting methods to facilitate the solutions of Maxwell's equations, the longest dimension of a system with respect to wavelength becomes the dominant characteristic. In a rectangular plane object x by y dimensions, the dimension affecting which technique is used will be the diagonal, L.

This relation connects the NEMR wavelength lambda to the diagonal L. The method used to solve Maxwell's equations would be:

1. Wavelength >> L, electric circuit theory (Kirchhoff's laws).
2. Wavelength = L, microwave theory or electromagnetic field theory or.
3. Wavelength << L, optics or ray theory.

In Kirchhoff's law in a circuit loop, all the voltages and currents add to zero, as shown in the equations in Eq. 2.13 A and B.

Equation 2.13: A and B Kirchhoff's Law

$$\sum_{\mathrm{n}}^{\mathrm{k=1}} \mathbf{I}_{\mathbf{K}} = \mathbf{0}$$

$$\sum_{\mathrm{n}}^{\mathrm{k=1}} \mathbf{V}_{\mathbf{K}} = \mathbf{0}$$

All the currents entering the circuit are equal to all those leaving a node. The node in an electrical circuit is a connection or junction of two or more parts, such as cables. Ray theory applies beyond the visible and ultraviolet light regions where particle theory is used.

Fields can interact with objects in the near or far field and give substantially different results. In the near field, field lines are not easily determined and are not always perpendicular, and B and H fields are characterized as waves. Far fields are where objects are placed at a distance from the source such that the *EM* waves are as from a point source. This means they are characterized as from a point source that is approximately a spherical wave and becomes, at further distances, a plane wave (advancing plane fronts rather than curved fronts), both being isolines, with the values of the field being the same along the front.

The absorption of energy by an object on the macroscale is a result of the field alignment with the largest dimension of the object and the material properties of the object in relation to what it interfaces. The permittivity, or dielectric constant, and the polarity or inducible polarity of the material determine this interaction. Although permittivity is referred to as a constant, it varies with frequency. It contains real and imaginary components. The real part is what is considered the vector that separates the charges in the material at a given set of conditions (frequency and *E* field strength). The complex part is related to the conductivity (specifically the conduction charge drift) of the material at a given frequency. These equations apply when the time-varying electromagnetic fields are described by *ejwt*, where $j = \sqrt{-1}$ and ω is the radian frequency (Eq. 2.14).

Equation 2.14: Permittivity

$$\varepsilon^* = \varepsilon_0 \left(\varepsilon' - \mathbf{j}\varepsilon''\right)$$

Again, ε_0 is permittivity in free space. The product of the real and imaginary part difference and the dielectric constant in free space equals the complex relative permittivity (ε^*). ε' is the dielectric constant, where ω is the radian frequency and *j* = the square root of −1. The time variation of the fields can also be described as in Eq. 2.15, where σ is the conductivity and frequency is expressed in radians (Eq. 2.16).

Equation 2.15: Complex Permittivity

$$\varepsilon'' = \sigma / \omega\varepsilon_0$$

Equation 2.16: Radial Frequency

$$\omega = \mathbf{2\pi f\ radians / s}$$

Then the effective dielectric constant (permittivity) is related to the sample's permittivity and the permittivity in free space within a given dispersion frequency range (*x*); *x* is a multiple of the frequency. The *s* subscript refers to the static condition (zero frequency). ε_∞ represents permittivity at infinite frequency (Eqs. 2.17 and 2.18).

Equations 2.17 and 2.18: Permittivity Based on Frequency and Conductivity Based on Frequency

$$\varepsilon = \varepsilon_\infty + \frac{\varepsilon_s - \varepsilon_\infty}{1 + x^2};\ \sigma = \sigma_s + \left(\sigma_\infty - \sigma_s\right)\frac{x^2}{1 + x^2}$$

Conductivity is related to the permittivity for a given frequency or range of RFR frequencies within a dispersion. Subscripts *s* and ∞ refer to static condition and infinite frequency, respectively.

The effective conductivity [6] has a similar relationship. Frequency, electric field, and conductivity are all tied together to result in the specific absorption rate (SAR) of the RFR or microwave (MW) energy (Eq. 2.19). All classical electromagnetic bioeffects are ascribed to SAR and subsequent heating, except for very low frequency, when electric shock comes into effect.

Equation 2.19: SAR

$$\mathrm{SAR} = \mathrm{P} / \rho_{\mathrm{m}} = \sigma |\mathrm{E}|^2 / \rho_{\mathrm{m}} = \omega \varepsilon_0 \, \varepsilon'' |\mathrm{E}|^2 / \rho_{\mathrm{m}}$$

ω is the radian frequency of the applied field. ε” is the measure of friction associated with changing polarization or charge drift. For magnetic materials, permeability is the effect of applied fields on the material and new fields created by moving magnetic dipoles in the material. P_m is the mass density of the object. P is the absorbed power.

The energy absorption resulting from the combined action of applied fields interacting with the intrinsic properties and dimensions of the material being irradiated is what produces the effects in which we are interested on a macro-measurable scale.

In short, the specific absorption rate (SAR) in watts/unit mass is higher when the incident *E* field is parallel to the largest dimension of the object and when the cross section of the object is perpendicular to the incident *H* field. The *K* parameter (the direction of NEMR propagation) is intermediate in SAR. The goal of NEMR dosimetry is to measure SAR that correlates with bioeffects (physiology) and, by extension, to adverse health effects (pathology) to determine safety exposure thresholds. At low frequencies approaching static fields, electric shorts (shock) become more important than SAR.

The general rules that affect NEMR interactions in addition to those discussed above with field orientation and object geometry are:

1. Sharp corners, points, and edges (high-aspect ratio geometries) concentrate *E* fields.
2. The uniformity of an incident field does not correspond to a uniform internal field.
3. The depth of penetration of SAR decreases at e^{-1} and as conductivity increases (*E* field = 0 below the surface of metal); as the frequency increases, conductivity in biological water-based material increases and the depth of penetration (skin layers) decreases.
4. Objects smaller than the wavelength of the NEMR cause little perturbation and/or scattering of NEMR fields (scattering is equivalent to external fields inducing internal fields).
5. The macroresonance of objects varies by an approximate factor of f^2 (square of the frequency).
6. At just below resonance, in *E* field polarization, SAR increases faster than f^2. Near beyond resonance, SAR decreases approximately by a factor of $1/f$ and becomes a steady state. Variations in SAR with frequency are most rapid near resonance.
7. Near and below the resonances, SAR is greatest for E polarization and least for H and intermediate for K polarizations.
8. For E polarization, the longer and thinner an object is, the greater the SAR.

The three measurements made for NEMR are E field, H field, and SAR. Measurements are made of temperature change and/or incident power density (by meter) to determine exposure and calculate dose. These are based on dipole antenna theory and various temperature-measuring modalities inserted into the exposed object, such as thermocouples or optical thermal sensing. The measuring device must minimally perturb the field, and the object being subject to the NEMR must have properties that reflect the biological physical responses.

Physical models to validate NEMR interaction predictions based on the previous fundamental principles and computations were first very simple, such as prolate spheroids. They were first filled with isotonic physiological saline to simulate tissue dielectric properties, but these were more for validating simplified mathematical models than representing real physical responses of biological tissues and whole bodies. The responses predicted by classical electromagnetic theory were based on dipole antenna theory and various temperature-measuring modalities inserted into the exposed object, such as thermocouples or optical thermal sensing. The measuring device must minimally perturb the field, and the object being subject to the NEMR must have properties that reflect the biological physical responses. As a result, tissue-equivalent materials were then composed of hydrophobic and hydrophilic materials, either homogeneously or prepared in layers with different dielectric properties. The shapes were made to duplicate animal or human body parts or whole individuals, made of NEMR transparent containment materials. Various nonperturbing temperature-measuring probes were inserted at various locations in the "phantoms" to measure the temperature changes from which local and average whole-body SARs could be calculated and normalized to incident radiation power densities for given frequencies and intensities.

Power (here in watts = joules/second) is generally defined as the amount of energy transferred or converted per unit time. These data confirmed the models (or falsified them) and became the basis for dosimetry related to real-life exposures and physiological and pathological responses extrapolated from animal experiments to human exposures.

Laboratory research on models to determine dosimetry with corresponding physiological responses of animals and field measurements of dosimetry under realistic exposures from RFR sources that led to current military and civilian international standards grew out of no small part the development of (1) 95-1-1997-IEEE Standard for Safety Levels with Respect to Human Exposure to Radio Frequency Electromagnetic Fields, 3 KHz to 300 GHz [4, 5]; and (2) the North Atlantic Treaty Organization (NAT0) Standardization Agreement (STANAG) 2345 MED (Edition 2) "Evaluation and Control of Personnel Exposure to Radio Frequency Fields—3KHz to 300 GHz," Military Agency for Standardization, October 13, 1997. Another important set of guidelines is the ICNIRP (International Commission on Nonionizing Radiation Protection) Guidelines on Limiting Exposure to Electromagnetic Fields for the protection of humans exposed to radiofrequency (RF) electromagnetic fields in the range of 100 kHz to 300 GHz. The guidelines include 5G technologies, WiFi, Bluetooth, mobile phones, and base stations. These guidelines replaced the 100 kHz to 300 GHz part of the ICNIRP (1998) radiofrequency guidelines and the 100 kHz to 10 MHz part of the ICNIRP (2010) low-frequency guidelines.

This stage brings us to the classical part of the paradox. The first requirement was to develop a molecular probe that would produce a signal that did not require intrusive instrumentation but could be precisely and accurately quantified with remote, already existing, instrumentation and correlate with the SAR (and/or the temperature change) of the NEMR absorption or the electric or magnetic fields. The molecular probe would have to be minimally perturbing to the fields in the material in which RFR absorption was being measured. If this probe could be incorporated into tissue and could measure internal energy absorption or fields, then it would be a plus. Initially, we thought the substrate would require a reduction/oxidation enzyme such as peroxidase or a type B cytochrome.

We linked an aromatic amino acid, 3-amino-L-tyrosine, to a luminescent reporting molecule, luminol (Fig. 2.4). The substrate was constructed not as a single binary molecule but as a polymer (inadvertently) (Fig. 2.5) [6].

The first step to making such a material that reports RFR and microwave absorption into surrounding material that correlates with temperature change was achieved with a long-term steady-state luminescent reaction extended to physiologic conditions. The first reaction to demonstrate this characteristic was catalyzed by horseradish peroxidase (HRP). However, it was then found not to require this enzyme for long-term thermally (RFR) sensitive steady-state emissions. Examples of these data are shown in Figs. 2.6 and 2.7.

They show the reaction with and without horseradish peroxidase using the substrate (diazoluminomelanin (DALM)) polymer at various pH values. Physiological pH (pH 7.4) gave the most closely matched HRP, and DALM alone responses at constant temperature. These responses were found to be of very long duration with very little material and still easily followed by measurement of light emission (measured out to 52 h).

In one instance, we set up a large rectangular prism (cuboidal) container in the lab in a chemical hood containing several liters of the luminescent reaction mixture

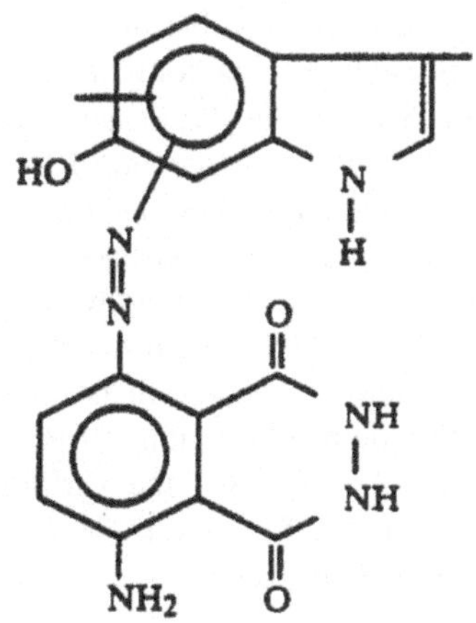

Fig. 2.4 Proposed synthetic structure of the luminescent peroxidase substrate that led to the discovery of diazoluminomelanin (DALM)

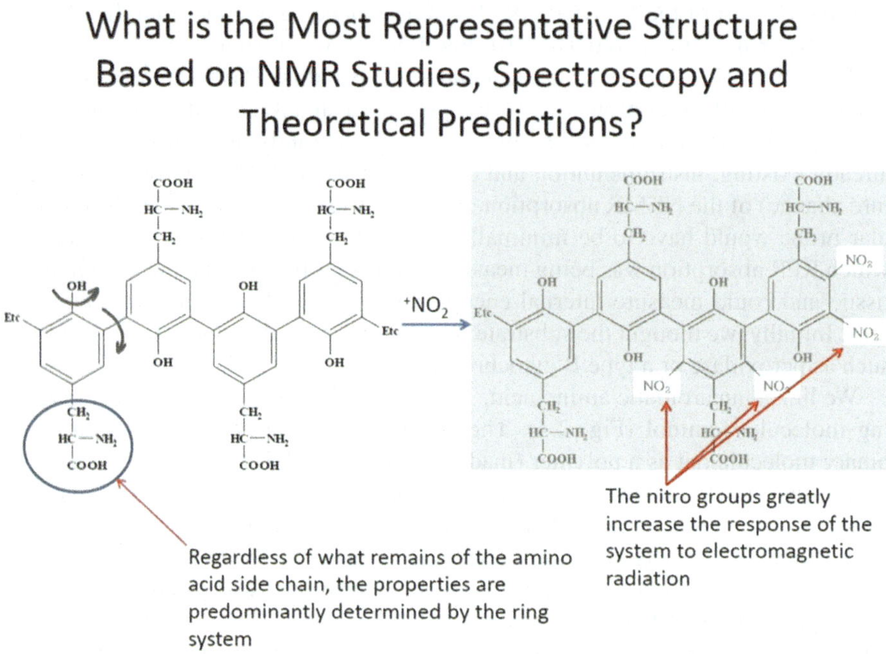

Fig. 2.5 The most likely structure of a polymer diazoluminomelanin

and observed it visibly in a darkened room (after adapting our eyes to the dark). The preparation luminesced for a whole year.

How was this reaction applied to classical RFR dosimetry and made compatible with existing theoretical models? Permittivity could not be used as a predictor. Because E fields are more easily determined at dielectric boundaries and then become a more complex mixture of reflected, refracted, and induced fields upon entering deeper into a material, the determination of the permittivity of complex structures is not very informative about the molecular level (the total results are difficult to deconvolve or theoretically deconstruct) [7]. Furthermore, in human and animal bodies, not only the complex geometries (they are not just prolate spheroids) but also the various tissue layers of the skin, muscle, organs, and fat lead to unusual fields and corresponding RF energy absorption patterns. Increased energy absorption, and subsequent heating, occurs at subcutaneous fat and fascial planes of muscle due to reflection back and loss at these dielectric interfaces (fat has a lower dielectric constant than water and skin). In particular, fibril (collagen) fat interfaces within fat layers are primarily responsible for local energy deposition and heating in this tissue. This result is further compounded by the variation in the penetration (skin depth) of various RFR frequencies with their respective accompanying intensities [8–14]. See the various editions of the Radiofrequency Radiation (RFR) Dosimetry Handbook for more detailed descriptions and formal explanations of these phenomena.

Fig. 2.6 Luminescence with and without horseradish peroxidase (HRP) with DALM and hydrogen peroxide at pH 6.5, pH 7.4, and pH 9.5, respectively. Note that at pH 7.4, there is no difference between HRP with DALM and DALM without HRP in delayed enhanced luminescence

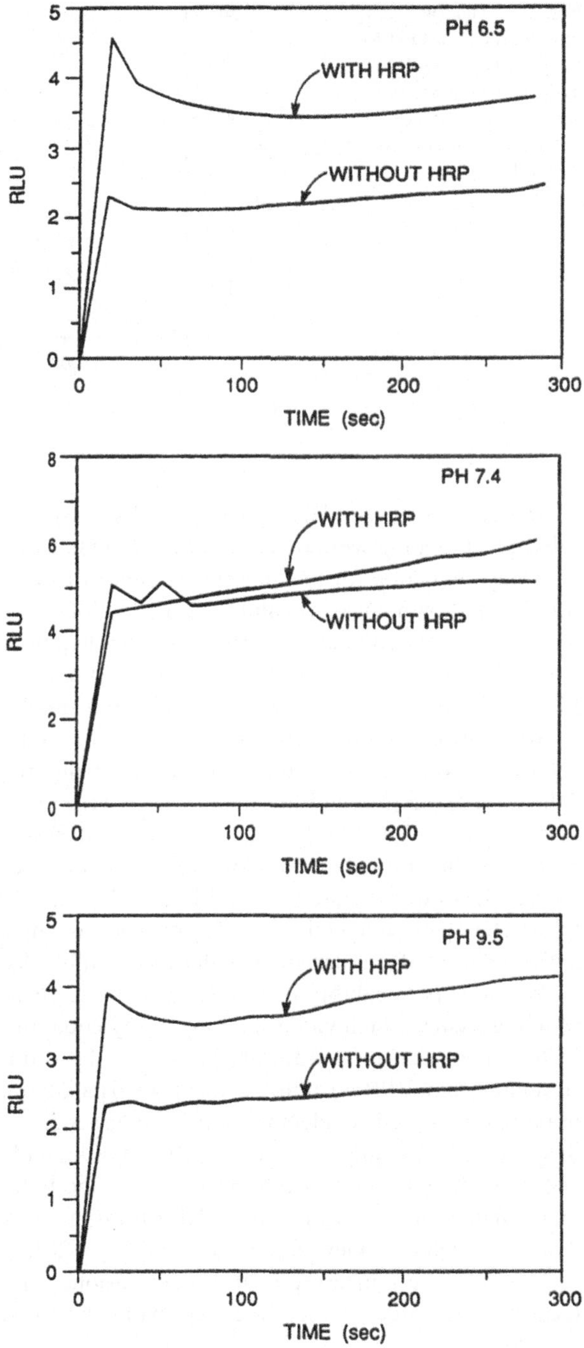

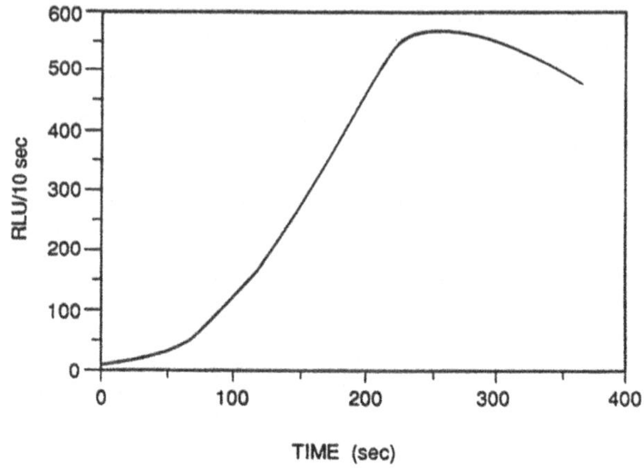

Fig. 2.7 The long-term luminescence of DALM alone activated with hydrogen peroxide in an aqueous solution (187.5 micrograms of DALM) of 0.56 mL

The attempts to model, computationally, tissue and organ RFR absorption and subsequent heating were built from block models of humans and animals based on the Debye equation for the electric influence of ions (Debye length or radius) in a solution and on membranes and the polarization and relaxation times following the cessation of applied fields (determining ion relaxation, dielectric constant, and complex permittivity). A finite difference time domain (FDTD) model is used to solve Maxwell's equations directly on a smaller and smaller block scale without physical approximation, and the problem of size is limited only by computing power. It determines Maxwell's equations on a mesh, computing E and H fields at grid points delta x, y, and z (three-dimensional (3D) Cartesian) apart. The E and H fields are determined in 3D. FDTD accounts for scattering, transmission, reflection, and absorption in a time domain. However, it can be extended to the frequency domain through fast Fourier transform (FTT) and discrete Fourier transform (DFT) operations. The electromagnetic field determinations impact various material properties at the surfaces of smaller and smaller resolved blocks determined by FDTD [15].

Magnetic permeability is of little importance because of the nonmagnetic nature of living tissue, which yields a permeability equal to that of free space.

Ion deposition on a different phase (liquid–solid interface) surface and nanostructures and RFR responses, these interfacial relationships will become very important in regards to electromagnetic interactions as we progress in this book. At this point, the assumption was that diazoluminomelanin (RFR-sensitive polymer) and its various formulations were reporting the bulk response in tissue equivalent material held in human and animal RFR transparent "phantoms" (Figs. 2.8, 2.9, and 2.10). The classical view appeared to hold [16–20].

However, these measurements, under various parameters, began to deviate from the expected values, which was not explained by standard errors or experimental errors.

Next, we demonstrated, under various RFR exposures, the reproducible reporting of energy absorption, which correlated with temperature simultaneously

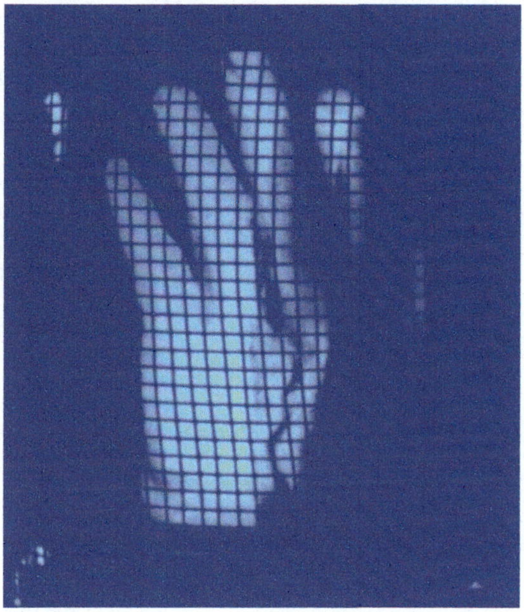

Fig. 2.8 A transparent human hand "phantom" filled with activated DALM solution luminescing in a 2450 MHz polarized radiation waveguide during microwave exposure

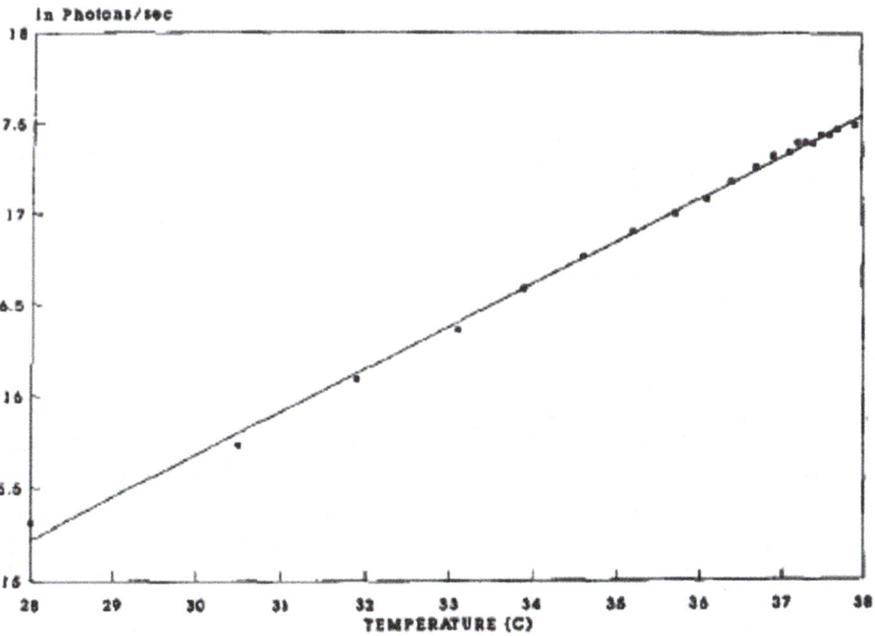

Fig. 2.9 Linear correlation of the log of luminescence with temperature

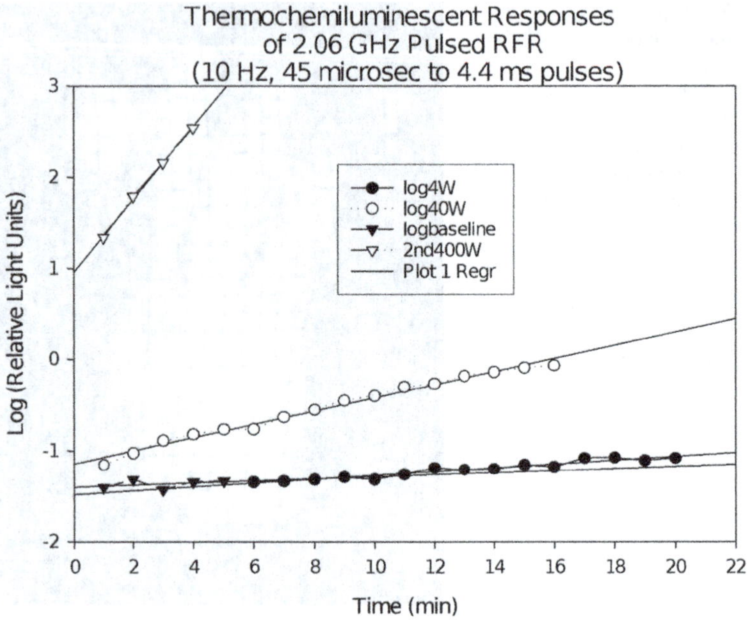

Fig. 2.10 Linear correlation of the log of luminescence with wattage (power) input of microwaves

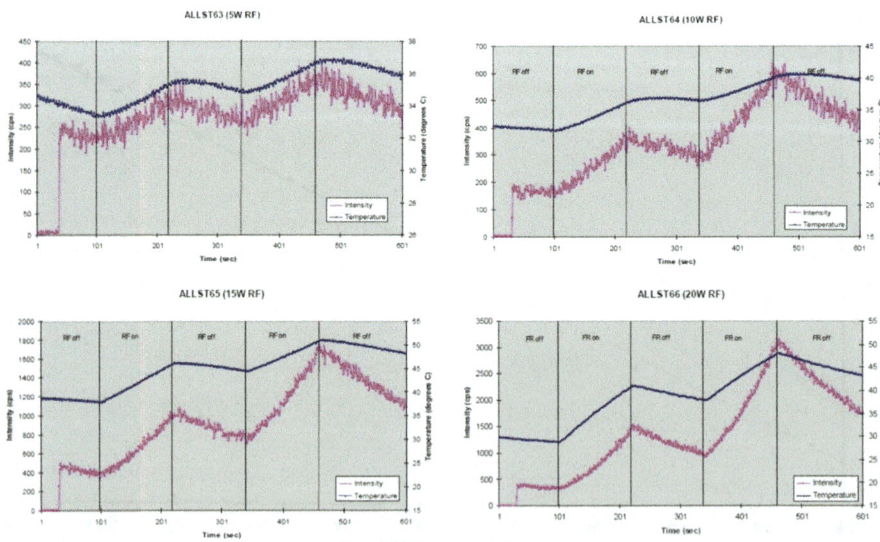

Fig. 2.11 Intermittent microwave exposures of activated DALM solutions, at various powers, in the 2450 MHz circularly polarized waveguide, showing thermal and luminescence responses

measured by other means. The best illustration and summary of the bridge between classic electromagnetic approaches to RFR dosimetry and "molecular dosimetry" promised by thermochemiluminescence (TCL) are shown in the paper by J. L. Kiel et al., Bioelectromagnetics 20:46–51, 1999 [17]. However, deviations (in circular waveguide exposure at various power densities) from expected TCL responses began to suggest mechanisms other than thermal (Fig. 2.11).

Another driver was the attempt to make the response in a solid state and recyclable to avoid the necessity of using a constructed chemical reaction for each repeated measurement. This approach (the "film badge" design) [21] demonstrated even more deviations from the classical electromagnetic mechanism (making the responses too complex to interpret into whole-body doses of RFR for health and safety monitoring). We end here and enter a new chapter of starting to characterize this anomalous behavior based on molecular structure and behavior in the electromagnetic fields of the DALM that may extrapolate to other biological molecules with common properties. Could these phenomena and paradoxes ever lead to useful applications?

References

1. Roach, W. P. (Editor). Radiofrequency Radiation Dosimetry Handbook (5th Edition), AFRL-RH-BR-TR-2010-0065, July 2009.
2. Durney, C. H., Massiudi, H., and Iskander, M. F. Radiofrequency Radiation Dosimetry Handbook (4th Edition), USAFSAM-TR-85-73, October 1986.
3. Wolski, A. Theory of electromagnetic fields, University of Liverpool, and the Cockcroft Institute, UK, pp. 15-65, https://cds.cern.ch/record/1400571/files/p15.pdf, June 2010.
4. IEEE Committee on Man and Radiation: TECHNICAL INFORMATION STATEMENT ON: Human Exposure to Microwaves and Other Radio Frequency Electromagnetic Fields, IEEE Engineering in Medicine and Biology Magazine 14(3):336–337, 1995.
5. IEEE International Committee on Electromagnetic Safety Technical Committee 95, Synopsis of IEEE Std C95.1TM-2019 "IEEE Standard for Safety Levels with Respect to Human Exposure to Electric, Magnetic, and Electromagnetic Fields, 0 Hz to 300 GHz", vol. 7, 171346–171356, 2019.
6. Kiel, J.L., O'Brien, G.J., Dillon, J., and Wright, J.R. Diazoluminomelanin: A Synthetic Luminescent Biopolymer. Free Radical Research Communications 8: 115–121, 1990.
7. Alon, L. et al., Effects of Anatomical Differences in Electromagnetic Fields, SAR, and Temperature Change, Conepts Magn Reson Part B, Magn Reson Eng 46(1): 8–18, Feb 1, 2016.
8. Cavagnaro, M. et al., From Magnetic Resonance Imaging to Dielectric Properties of Tissues, Biophysics & Bioeng. Letters 4(2): 1–8, 2011.
9. Costanzo, S., et al. Gel-like Human Mimicking Phantoms: Realization Procedure, Dielectric Characterization and Experimental Validations on Microwave Wearable Body Sensors, Biosensors 11: 111, 2021.
10. Baker-Jarvis, J., and Kim, S., The interaction of Radio-Frequency Fields with Dielectric Materials at Macroscopic to Mesoscopic Scales, J Res NIST 117:1–60, 2012.
11. Nasir, N., and Ahmad, M., Review Article: Cells Electrical Characterization: Dielectric Properties, Mixture, and Modeling Theories, Hindawi J of Engineering, 1–17, 2020.
12. Hirata, A., et al., Human exposure to radiofrequency energy above 6 GHz: review of computational dosimetry studies, Phys. Med. Biol. 66, 08TR01, 2021.

13. Wang, P., and Anderko, A., Computation of dielectric constants of solvent mixtures and electrolyte solutions, Fluid Phase Equilibria 186: 103–122, 2001.
14. Adair, E., and Petersen, R., Biological Effects of Radiofrequency/Microwave Radiation, IEEE TRANSACTIONS ON MICROWAVE THEORY AND TECHNIQUES 50, (3): 953–962, March 2002.
15. Schneider J., Understanding the Finite-Difference Time-Domain Method, May 28, 2021, https://eecs.wsu.edu/~schneidj/ufdtd/ufdtd.pdf.
16. Kiel, J., Alls, J., Holwitt, E., Stribling, L, and Parker, J., Thermochemiluminescence as a technique for radio frequency radiation dosimetry. Bioelectrochemistry and Bioenergetics 47: 253–257, 1998.
17. Kiel, J., Alls, J.., Mason, P., and Erwin, D., Luminescent Radio Frequency Radiation Dosimetry. Bioelectromagnetics 20: 46–51, 1999.
18. Kiel, J., Molecular Dosimetry. In Radio Frequency Radiation Dosimetry (B. J. Klauenberg and D. Miklavcic, eds.), Kluwer Academic Publishers, Netherlands, pp. 227–237, 2000.
19. Erwin, D., Batishko, C., Stahl, K., and Kiel, J. (AF Inv 17,590) Quantitative Luminescence Imaging System. U.S. Patent 4,948,975 (issued 14 Aug 1990).
20. Kiel, J.L., and O'Brien, G.J. (AF Inv 18,422) Luminescent Polymer. U.S. Patent 5,003,050 (issued 26 March 1991).
21. Holwitt, E., Kiel, J., and Erwin, D. (AF Inv 20884) Microwave-sensitive Material. U.S. Patent 5,658,673 (issued 19 Aug 1997).

Chapter 3
Theory and Measurement: Relaxation Versus Resonance

Abstract Could classical electromagnetic mechanisms be extended to molecules from the macrostructures they compose? Can the top-down approach be used? In this chapter, we will examine that premise to exhaustion and see what remains. We will end with the possible mechanisms that have been suggested and examined beyond thermal mechanisms and assess whether these are fundamental and how they may be dependent on some aspects of classical theory or whether they necessitate new theories.

Keywords Relaxation · Resonance · DNA · Heme · Nitric oxide · Activation energy · Vibrations · Type-B cytochromes · Hemoglobin

In examining tissue and whole-body differences in dielectric constants, over a range of frequencies, several structural and molecular components become dominant over others over a large range. These dielectric constants are strongly affected not only by frequency but also by temperature. Muscle was used as a representative tissue. The dominant effects on dispersion are designated as alpha, beta, and gamma.

Dispersions are drops in the dielectric constants with small changes in frequency at relaxation frequencies for a given structure or component. Alpha dispersion is for membrane properties; for beta, tissue interfaces (Maxwell-Wagner effect); and for gamma, the water molecules' polarity (Debye effect) of the fixed water dipoles and electrolytes (and, to a much lesser extent, biological molecules such as proteins, amino acids, and soluble carbohydrates). Gamma dispersion accounts for most of the tissue-bound water effects on the dielectric constant equivalent to free aqueous solutions of these components. This last property is why physiological saline is used as a physical model for tissue. These dependencies have been described by Schwann, 1975 [1]. The dielectric constants (epsilon) and conductivity (sigma) can be formally described by Eqs. 2.17 and 2.18.

As described earlier, x is a multiplier of frequency for the dielectric constant determined by the beginning and end of the dispersion. ε_s is the static dielectric constant, and σ_s is the static conductivity.

J. L. Kiel, *Bionanotechnology*, https://doi.org/10.1007/978-3-031-63006-4_3

ε_∞ and σ_∞ are the same as their respective values in free space. The tissue water properties are determined by the conductance of ions in water, the relaxation frequency of water, and the static dielectric permittivity at frequencies much less than the relaxation frequency of pure water at 25 °C, that is, 20 GHz. The relaxation frequency of pure water at 37 °C (body temperature) is 25 GHz. At 1 °C, the relaxation frequency of water is 9 GHz. At room temperature, the relaxation frequency of tissue water is 20 GHz, similar to pure water. Static and infinite frequency permittivity values are 78 and 5, respectively, at room temperature (25 °C).

The conductivities, based on the calculations, should be 0.8 mho/cm at frequencies above 20 GHz at 25 °C. This value is much larger than the typical low-frequency conductivities of biological liquids at 0.01 mho/cm. The dielectric properties are unaffected by the field strength until it exceeds 100 kV/cm. The electrolyte solutions have almost identical dielectric static behavior as water. Static conductivity depends on the conduction of ions added. The static dielectric permittivity of water with electrolytes at physiological concentrations (0.15 normal) is 2 units lower than that of pure water (a small difference). Therefore, the dielectric dispersions over a large frequency range should indicate the components of a complex biological solution or heterogenous phase preparation. These assumptions can be extended to complex structures such as polymeric colloids and nanoparticles, including covalent and noncovalent (ionic layers) components.

This review of human (and extrapolation to and from experimental animals and models) radiofrequency radiation (RFR) dosimetry is necessary for three reasons. First, it determines safe levels of exposure. Second, it determines if these levels correspond to activation levels for the in vivo application of radio-controlled antimicrobial and genetic vectors. Third, this information is necessary to determine what tissues and organs would reach desirable levels of E and H fields with such levels of RFR exposure to activate bionanotechnological applications.

The mechanisms that come into play in the frequency range from static to megahertz, including human body resonance, 80–100 MHz, with the average man at 70 MHz, are displayed in Table 3.1, based on Schwann, 1975 [1].

Biological molecule interactions with RFR lead to the classical electromagnetic analysis of the DALM polymer and provide guidance as to what to expect from certain structural designs of biological molecules, that is, what molecular characteristics optimize RFR interaction (at least from a classical sense). Bulk responses of the DALM polymer correlated, in a physiologic solution, with the thermography (infrared (IR) camera) frames of transparent plastic rat phantoms at various frequencies of exposure. Thermography is a surface measurement, whereas thermochemiluminescence (TCL) is the sum of all emissions within the exposed volume in a given line of sight. The latter is why the thermography of millimeter wave (with a very shallow level of penetration) exposure is more intense than the corresponding TCL.

The frequencies were chosen to reveal different depths of penetration and illustrate response times that are too fast to be confounded by fluid convection or thermal diffusion (Fig. 3.1).

Table 3.1 Dielectric relaxation and dispersion properties of tissues, cells, organelles, and molecular components

Components	Relaxation and dielectric dispersions
Electrolytes	Gamma[a,b]
Amino acids	Sigma[c] and gamma
Proteins	Beta[b], sigma, and gamma
Nucleic acids	Alpha[d], beta, sigma, and gamma
Cells in general	Beta and gamma
Charged cells	Alpha, beta, and gamma
Excitable cells	Alpha, beta, and gamma

[a]Permanent-dipole rotation (Debye)
[b]Inhomogeneous structure (Maxwell-Wagner)
[c]Bound water, side chains of amino acids, rotations, and fine structural effects
[d]Subcellular organelles (Maxwell-Wagner) and counter-ion relaxation

However, the bulk properties of responses and the measures of various DALM polymer preparations began to reveal anomalies that deviate from classical responses (Figs. 3.2, 3.3, 3.4, 3.5, and 3.6).

The dielectric properties of biosynthetic DALM showed that it did not significantly alter the permittivity of water, even though it demonstrated the same luminescence and microwave response as synthetic DALM.

The frequencies where each dispersion is dominant are alpha, $1–10^4$ Hz; beta, 10^4 Hz to 10^8 Hz (10 KHz to 100 MHz); delta, 100 MHz to 1 GHz (10^9 Hz); and gamma, 20 GHz (2×10^{10} Hz). The dispersion frequencies are where the various components absorb the most energy. Some of these are most likely to interact with nanostructures, and others with biological or other molecules and ions. Together, these interactions could account for nanoparticle and molecular effects or a combination thereof.

However, over the years, there have been various research efforts to find specific radiofrequency (RF) molecular resonances associated with coupling to whole or subunit molecular structures and their natural frequencies, as with heme and deoxyribonucleic acid (DNA) [2–5]. These biological molecules, because of their importance, have been the drivers for these hypothetical pursuits. The question we pursued is: Is there any phenomenological evidence of a biological molecule specifically absorbing RF independent of the relaxation frequency and dispersion mechanism? If so, what plausible mechanism could be engaged? Are there any other biological molecules that possess the same or similar properties that could extend the mechanism beyond a specialized molecule, such as DALM, to create more generalized biological, physiological, or pathological effects? The data I now present provide hints to the answers to these questions.

E. W. Prohafsky reviewed the theoretical basis for molecular interactions based on the fundamental intramolecular frequencies of biological molecules (proteins) through the measurement of acoustic velocities in the bulk solutions of those molecules [6]. The lowest frequencies of intramolecular resonance absorption would be multiples of hundreds of GHz. Any lower frequencies would be determined by

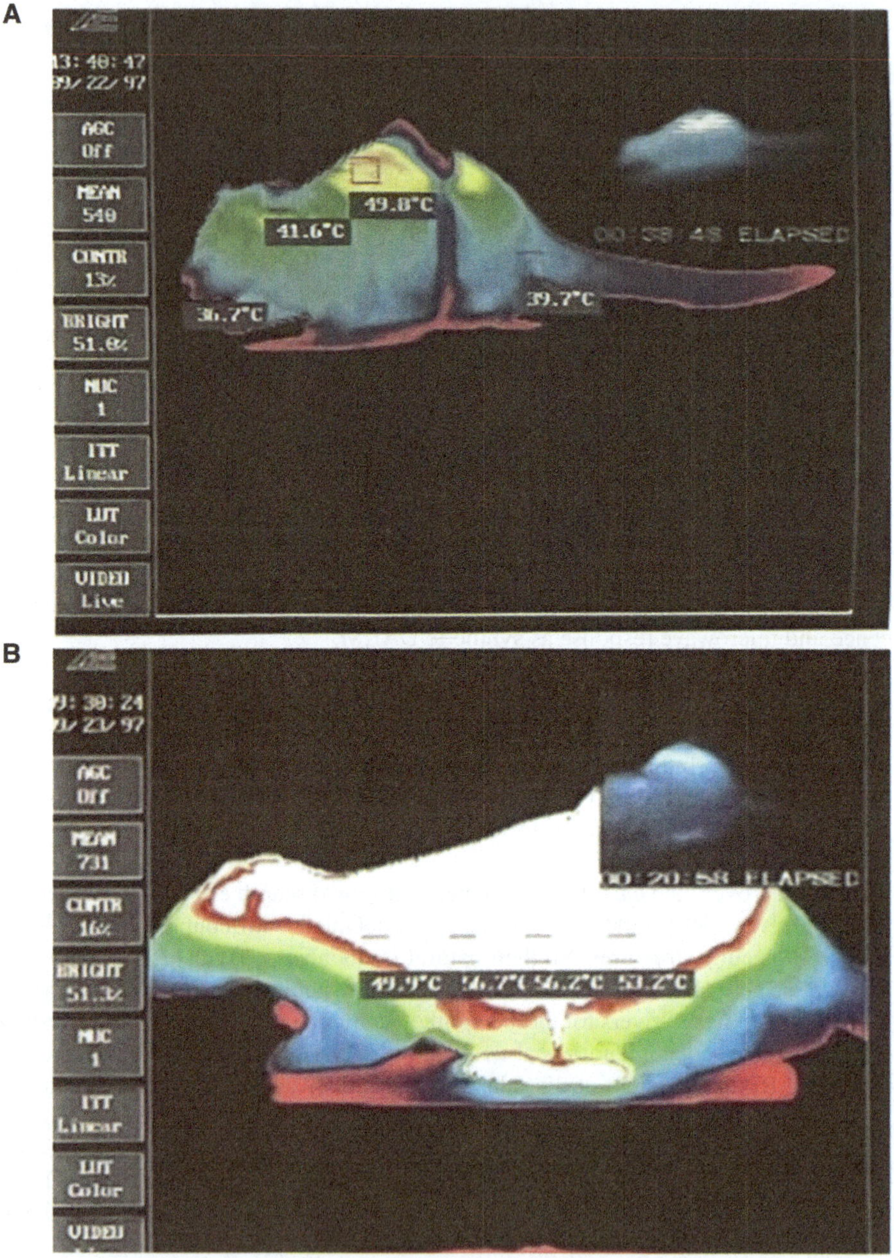

Fig. 3.1 Thermography compared simultaneously to the thermochemiluminescence of the DALM solution in rat phantoms at 2.06 MHz and 35 MHz, respectively. Infrared video image (pseudocolorized) showing thermal distributional data (temperatures of discrete regions (squares) given) and visible low-light image (inset) showing the true color (blue) of thermochemiluminescence (TCL) distribution in a rat phantom for comparison. The first rat phantom, filled with a diazoluminomelanin-containing solution, was exposed to 2.06 GHz for a period of 18 min, 5 s. The images are single-

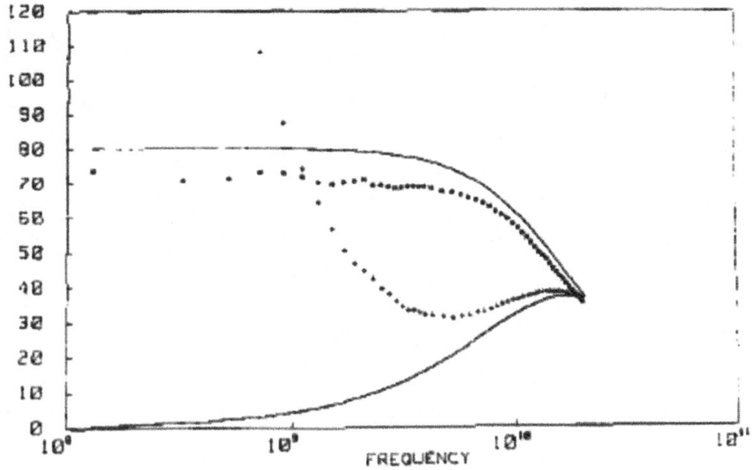

Fig. 3.2 Dielectric properties of the aqueous solution of diazoluminomelanin (DALM). The vertical axis is the unitless dipole moment (permittivity). The dielectric properties of diazoluminomelanin (DALM) in water compared to water alone measured with impedance and network analyzers. The upper continuous line (epsilon prime) is the real component of the permittivity of water, and the lower continuous line (epsilon double prime) is the complex part of the permittivity (loss). The stars and crosses represent the values of epsilon prime and epsilon double prime, respectively, for the aqueous solution of DALM

Fig. 3.3 Luminescence with respect to the time of exposure and the temperature of various DALM liquid preparations compared at an exposure of 2450 MHz, 25 W, continuous wave in a circularly polarized waveguide

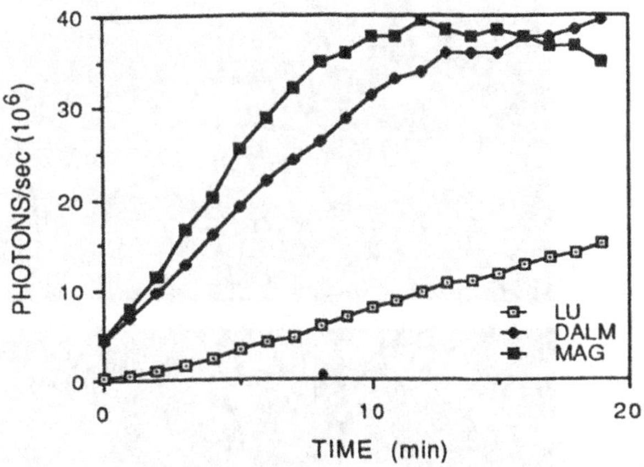

Fig. 3.1 (continued) captured frames closely matched in time. The infrared image is a single video frame captured from a group recorded at 30 frames per second. The low-light image is a frame summation of 2 min of TCL. The dark broken streak down the side of the rat phantom is a cooling artifact from the TCL liquid that flowed out of the fill port in the top of the phantom because of foaming caused by the reaction. The second image is of infrared and low-light (inset) video images of a rat phantom filled with thermochemiluminescent fluid and exposed to 35 GHz at 1 W/cm^2, the average incident power, for 10 min, 58 s

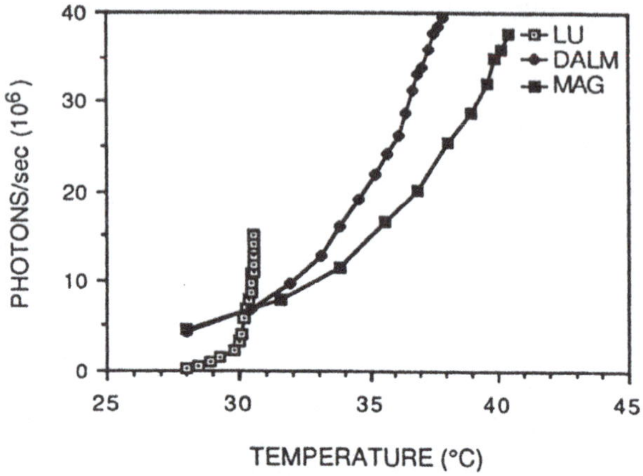

Fig. 3.4 Luminescence with respect to the temperature of various DALM liquid preparations compared at an exposure of 2450 MHz, 25 W, continuous wave in a circularly polarized waveguide

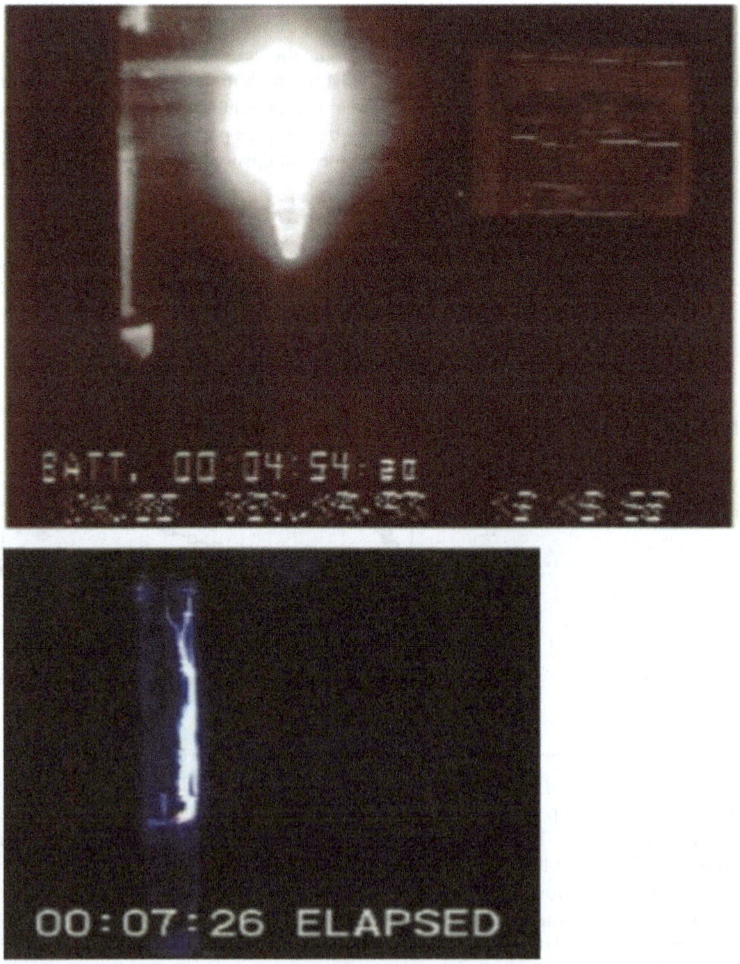

Fig. 3.5 1.25 GHz, 10 pps, 6 us pulse width, 2 MW peak power; the pulse responses of activated DALM solution

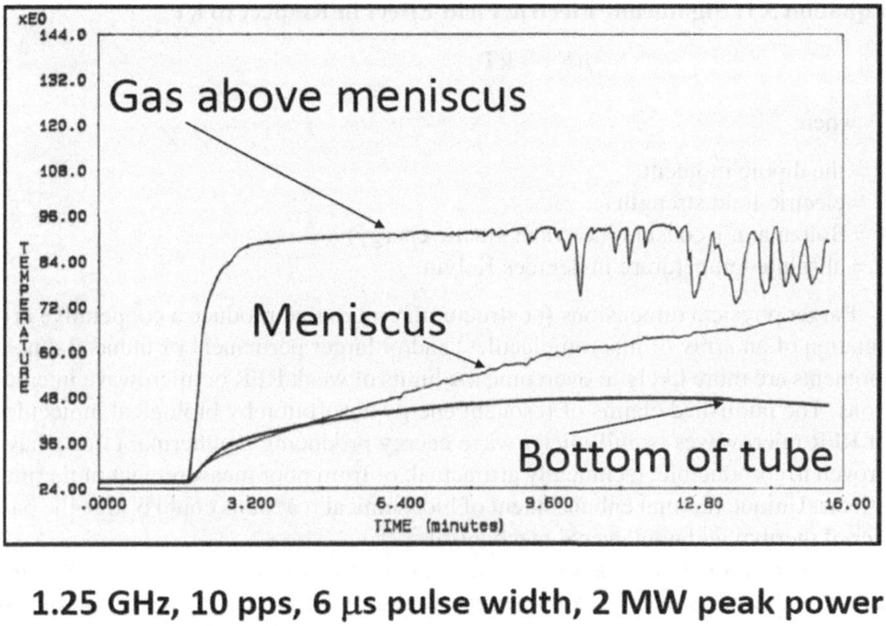

Fig. 3.6 Temperature distribution in a tube of activated DALM solution during exposure to 1.25 GHz, 10 pulses per second, with a 6-microsecond pulse width, at 2 MW peak power

intermolecular interactions. These latter frequencies were described earlier based on mechanisms outlined by H. Schwann. Even DNA lower frequency modes have been determined to result from the bulk effects of water interactions with the DNA. He determined that only frequencies above 4 GHz could result from DNA intrahelical modes, rather than from the DNA double helix coupled to the surrounding material. These 4 GHz absorbances are heavily dispersed and therefore cannot demonstrate resonance absorption. The result of his analysis supports the classical electromagnetic mechanism of the thermalization of bulk biological materials with no nonthermal resonant energy absorption. There have been a number of bioeffect studies since these conclusions purported to support nonthermal DNA resonance absorption, but direct physical evidence of this mechanism is lacking.

The relaxation absorption effects on biological macromolecules in a solution or tissue are dependent on their ability to follow the changes in fields in alignment. A few kilowatts per cm are required to align these long biological macromolecules and even higher fields to excite internal vibrations or produce submolecular orientation. Additionally, local fields yield very large barriers (lipid bilayer of cell membranes, with accompanying surface counter ions, near breakdown voltage, approaching 10^6 volts/meter).

These fields range from 1.5 kV/cm at 10 nm from monovalent ions to 1.8 kV/cm at similar distances from a heme-containing protein such as hemoglobin, myoglobin, peroxidase, or other type-B cytochromes. Molecules can be significantly reoriented if the dipoles and E-field strength are much greater than kT (Eq. 3.1).

Equation 3.1: Significant Electric Field Effect in Respect to kT

$$\mu \mathbf{E} >> \mathbf{kT}$$

where

μ = the dipole moment
E = electric field strength
k = Boltzmann's constant (average kinetic energy)
T = absolute temperature in degrees Kelvin

Larger physical dimensions (or structural bridges that produce a cooperative orientation of an array of macromolecules) and/or larger permanent or induced dipole moments are more likely to overcome the limits of weak RFR or microwave interactions. The published claims of resonant energy absorption by biological molecules of RFR microwaves or millimeter wave energy producing nonthermal effects have proven irreproducible, technically artifactual, or from poor measurement of thermal effects. Unique thermal enhancement of biochemical reactions could bridge the barrier of thermal and nonthermal mechanisms [7].

The deoxygenation or autoxidation of hemoglobin has proven to be a thermal effect, with microwave radiation being linked only to deviations in existing hyperthermal mechanisms (change in the activation energy of thermally enhanced autoxidation) [8–10].

Action on enzymes such as nitric oxide synthase (a type-B cytochrome and flavin adenine dinucleotide (FAD) flavoprotein) in situ in mouse macrophages has shown enhancement with low specific absorption rate (SAR)/high E-field ultrawideband radiation (indeterminant frequency based on very short pulses, picoseconds, which are subject to the uncertainty principle due to the pulse being shorter than a wavelength). This enhancement of the enzymatic-mediated reaction rate was dependent on biochemical stimulation of the enzyme but disappeared with genetic manipulation producing high constitutive levels of enzyme activity [11].

The first use of peroxidase with a luminescent substrate was thought to presumably detect the enhancement of the NEMR stimulation of activity because of the protoporphyrin IX heme content (a circular closed circuit formed from the organic semiconductor model). This anticipated result never materialized, with no enhanced properties resembling a semiconductor. The next step was to find or synthesize a larger, more linear "circuit" to interact with RFR and microwave fields.

Soon after its synthesis on March 1, 1988, DALM, in the summer of 1988, appeared to demonstrate such NEMR sensitivity. However, to avoid assuming molecular resonance without definitive evidence, we asked ourselves whether it was based on classical electromagnetics, quantum effects, or both or neither. Could we falsify any of these hypotheses?

The first dielectric measurements of pure aqueous solutions of the DALM polymer showed that complex permittivity had a high imaginary component at low frequencies (higher low-frequency conductivity) and a minor depression in the values of the real component (dielectric constant) equivalent to water containing a diluted colloidal polymer [12]. These results were not anticipated because the large polymer fibers of DALM (observed under a dark field in a light microscope) were thought to become partially oxidized, creating trapped ionic unpaired electrons along the length, which would, in turn, create a capacitance by the further separation of these charged radicals by the applied E field. The dielectric properties of the polymer at various frequencies were that of an inductor, not a capacitor. The sharp, high-aspect-ratio structure at the end of the polymer fibers, or the ribbons formed when the fibers were stacked in a staggered fashion along the length, should have concentrated applied E fields at the ends. This "lightning rod" effect has been observed with carbon nanotubes when RFR was applied, allowing them to penetrate the bacterial cell walls [13–15].

This effect was still a classical electromagnetic one. Furthermore, the thermalization response (steady state with fixed temperature) was highly linear when the natural log of the photons per second emissions were plotted against the temperature [10, 16]. These data also supported a thermal mechanism of electromagnetic absorption into the material. However, deviations in expected behavior began to reveal themselves in the thermochemiluminescence of the polymer components and various formulations of the DALM under the same conditions of NEMR exposure. Notably, the steep thermal threshold of luminol luminescence and the greater luminescence of iron-particle-bound DALM at the time of microwave exposure, but at higher temperatures than with luminol or free DALM luminescence alone, were some of the anomalies [16]. Free DALM luminesced less than iron-bound DALM over the time of exposure but showed a higher temperature sensitivity than iron-bound DALM. The greatest deviations were seen in discontinuous or high-pulsed power microwave exposures.

Continuous wave (2450 MHz) exposure in a circularly polarized waveguide (for more uniform exposures) produced heating following a predictable course with which luminescence, measured simultaneously, correlated. The exposures were turned on and off manually at various powers up to 20 watts. The accumulation of heating was as expected. However, at exposures to higher incident microwave power, the thermochemiluminescence began to deviate from the linear, parallel and correlated path from that of thermalization. With added heating at higher power, the thermochemiluminescence began to become exponential.

When high-power pulsed microwaves were used, the steady-state luminescence was enhanced by intense flashes that correlated with the microwave pulses. To rule out boiling and luminescent gas release, we instrumented the exposure tubes containing the reaction solution with thermocouples and collected data that showed that the bulk solution was cooler than the meniscus and the gas that was released at the surface. Flashing was not a result of overheating or boiling because it disappeared when the same bulk solution, lacking only DALM, was substituted in the exposure device. Thus, these phenomena gave birth to the paradox that extreme luminescence could be uncoupled from the heating (temperature change or average kinetic energy, the Maxwell-Boltzmann distribution) response to NEMR absorption in the bulk solution.

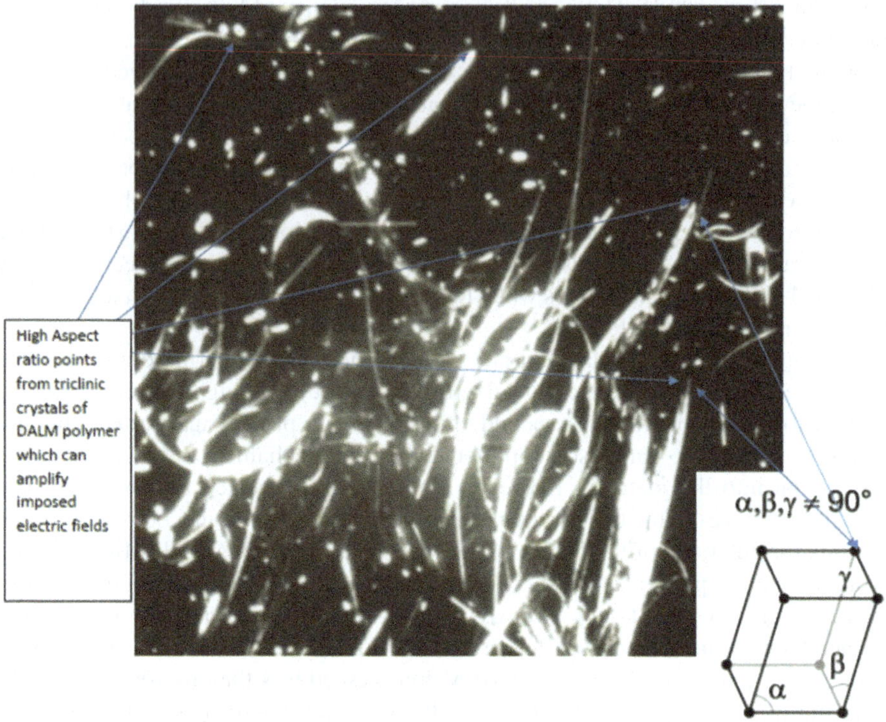

Fig. 3.7 The ribbon and angular high-aspect end structures of the synthetic DALM are probably due to the triclinic crystalline basic structure of the DALM polymer and the staggered alignment of the twisted (rotated) linear polymer molecules

One explanation, confirmed by the spectral measurements of the luminescence, was that the gas rose to the meniscus and surface of the solution, where it carried away the superheated electrons and ions of gas (approaching 5000–20,000 K, or 8540–35,540 F) and cooled extraordinarily quickly but still left some measurable increase in temperature, exceeding that of the solution. The bottom of the tube, where the bulk of microwave absorption should have occurred based on geometry and classical electromagnetic dynamic theory, cooled (contrary to the classical second law of thermodynamics).

The physical structure of DALM, revealed by microscopic examination, suggested a nanotechnological but classical explanation for the pulsed RF behavior (Fig. 3.7).

In 2006, a proposal was made for the same mechanism of nanotube inductors and their probable highly advantageous behavior used in microelectronics to be extended to DALM (fibers) and even be duplicated by either carbon nanotubes or DALM interchangeably [17]. Carbon nanotubes and DALM fibers are nanometers in diameter, so the magnetic field current in these structures is on the order of 1000 times that of a copper wire several microns in diameter. The magnetic field H is proportional to the radius and Z, the distance of the measuring cantilever from a surface (Eq. 3.2).

Equation 3.2: Nanotube Magnetic Field Current

$$\mathbf{H}_{\max}(\mathbf{Z}) = \mathbf{H}(\mathbf{r},\mathbf{Z}) = (\mu_0 \mathbf{I} / 2\pi)\left(\mathbf{r} / \left(\mathbf{r}^2 + \mathbf{Z}^2\right)\right)$$

The relationship between the magnetic field H and the inductance L of the nanoinductance is because the large magnetic field in these nanoinductors results in a large inductance (Eq. 3.3).

Equation 3.3: Nanotube Inductance

$$\int \left(\mu_0 \mathbf{H}^2 / 2\right) \mathbf{dV} \approx \mathbf{LI}^2 / 2$$

These inductors store energy in electron magnetic fields, thus slowing down surges and any changes in the current. This storage would explain the buildup of flashes and the storage of energy in DALM, driving luminescence. The current at lower frequencies passes through the nanoinductors, and high-frequency energies (especially pulsed) are stored momentarily in the nanoinductors. Thus, these inductors act as low-pass filters. Macro-inductors are cone shaped, as are the high-aspect ends of carbon nanotubes and DALM fibers. The conical shape limits stray capacitance and acts as a series of narrow-band inductors that create impedance (AC resistance) over a wide frequency bandwidth. When DALM was produced biosynthetically in bacteria, the dielectric characteristics previously described disappeared in the measurements, but the pulse phenomena persisted. How could this be? The question will be addressed in the next chapter.

References

1. Schwan, H., Dielectric properties of biological materials and interaction with microwave fields at the cellular and molecular level. In S. Michaelson and M. Miller (eds.), Fundamental and Applied Aspects of Nonionizing Radiation., New York: Plenum Publishing Co., 1975.
2. Panagopoulos D., Johansson O., and Carlo G., Evaluation of Specific Absorption Rate as a Dosimetric Quantity for Electromagnetic Fields Bioeffects. PLoS ONE 8(6): e62663. doi:https://doi.org/10.1371/journal.pone.0062663, 2013.
3. Mahamoud Y, et al., Additive Effects of Millimeter Waves and 2- Deoxyglucose Co-Exposure on the Human Keratinocyte Transcriptome. PLoS ONE 11(8): e0160810. doi:https://doi.org/10.1371/journal.pone.0160810, 2016.
4. Lawler, N., et al., Millimeter waves alter DNA secondary structures and modulate the transcriptome in human fibroblasts, Biomedical Optics Express: 13 (5),3131, 1 May 2022.
5. Gabriel, C., et al., Dielectric behavior of aqueous solutions of plasmid DNA at microwave frequencies, Biophys. J. 55: 29–34, 1989.
6. Prohofsky, E., RF absorption involving biological macromolecules, Bioelectromagnetics 25 (6): 441–451, 2004.

7. Dudley, G., Richert, R., Stiegman, A., On the existence of and mechanism for microwave-specific reaction rate enhancement, Chem.Sci. 6, 2144–2152, 2015.
8. Kiel, J.L., and Erwin, D.N. Microwave and Thermal Interactions with Oxidative Hemolysis. Physiological Chemistry and Physics and Medical NMR 16: 317–323, 1984.
9. Kiel, J.L., and Erwin, D.N. Microwave Effects on the Thermally Driven Oxidase of Erythrocytes. International Journal of Hyperthermia 2: 201–212, 1986.
10. Kiel, J.., McQueen, C., and Erwin, D., Interactions of Methemoglobin and Green Hemoprotein in Chemiluminescent Gels, Free Radical Biology and Medicine 8: 127–131, 1990.
11. Seaman, R., Parker, J., Kiel, J. Mathur, S.., Grubbs, T., and Prol, H., Ultra-Wideband Pulses Increase Nitric Oxide Production by RAW 264.7 Macrophages Incubated in Nitrate, Bioelectromagnetics 23: 83–87, 2002.
12. Kiel, J., Gabriel, C., Simmons, D., Erwin, D., and Grant, E., Diazoluminomelanin: A Conductive Luminescent Polymer with Microwave and Radiowave Absorptive Properties. In Proceedings of the Twelfth Annual International Conference of the IEEE Engineering in Medicine and Biology Society, (P.C. Pedersen and B. Onaral, eds.), Vol. 12, No. 4, IEEE, Philadelphia, 1689–1690,1990.
13. Giersig, M., et al., Novel Electroporation System for both Gram-negative and Gram-positive Bacteria Assisted by Multi-Walled Carbon Nanotubes, Mater. Res. Soc. Symp. Proc. 845: 285–290, Materials Research Society, AA5.21, 2005.
14. Atiyah, A., Haide, A., Dhahi, R., Cytotoxicity properties of functionalised carbon nanotubes on pathogenic bacteria, IET Nanobiotechnol 13 (6): 597–601, 2019.
15. Gao, J., et al., Nanotube assisted microwave electroporation for single cell pathogen identification and antimicrobial susceptibility testing, Nanomedicine: Nanotechnology, Biology, and Medicine 17: 246–253, 2019.
16. Kiel, J.L., O'Brien, G.J., Simmons, D.M., and Erwin, D.N. Diazoluminomelanin: A Synthetic Electron and Nonradiative Transfer Biopolymer. In Charge and Field Effects in Biosystems-2, (M.J. Allen, S.F. Cleary, and F.M. Hawkridge, eds.), Plenum Press, New York, pp. 293–300,1989.
17. Tsubaki, K., et al., Proposal of Carbon Nanotube Inductors, Journal of Physics: Conference Series 38: 49–52, 2006.

Chapter 4
Phenomenology of RFR Absorption by Biological Molecules and Membranes

Abstract The search for nonthermal-specific molecular resonances has been a search for "ghosts." The field has been plagued with irreproducible results and non-dose responsive effects ("windows" of responsiveness). Furthermore, the nonuniformity of radiofrequency radiation (RFR) absorption and the large variation in this property across the frequency range have led to the poor measurement of energy absorption indicated by temperature changes (classical electromagnetic mechanisms). One might say the precision and appropriateness of measurement are everything in RFR bioeffects. A central aspect of the problem is trivializing the importance of dosimetry by biologists, as well as engineers and physicists trivializing the difficulty of making such measurements in biological preparations in situ during RFR exposure.

Keywords Molecular resonances · Dipoles · Acoustics · Mitochondria · Chloroplasts · Electron transport · Membranes · NADPH oxidase

The theoretical approaches for explaining purported molecular-specific interactions with radiofrequency radiation (RFR) are colored by phenomenology. This "filter" is a set of assumptions that lead to hypotheses which data are selected to support. This approach sometimes led to less than careful measurement to avoid artifacts or the inappropriate comparison of equilibrium heating to RFR heating. Conventional heating and RFR and microwave heating were not comparable under these conditions. The difference was trivialized.

Observed differences in the effects of RFR heating and conventional heating led to the assumption that weak fields affect molecular or cellular to subcellular structures. This assumption revolves around the formation of cooperativity, that is, the formation of a large dipole moment across enzymatically active or electron transport systems. Such systems include mitochondria chloroplasts and shorter chains such as those of nicotinamide adenine dinucleotide phosphate hydride (NADPH) oxidase and nitric oxide synthase.

Froelich, as early as 1973 and 1975, suggested that such giant dipoles, composed of cooperative enzyme pathways, over many cells, could be responsible for such

J. L. Kiel, *Bionanotechnology*, https://doi.org/10.1007/978-3-031-63006-4_4

resonance behavior [1, 2]. This resonance behavior was found and modeled in the kinetics of the enzymes working as a network. Membranes, because they, in some cases, physically constrain enzyme systems in a single membrane or layers of connected membranes in a capacitively charged environment, were assumed to be the most likely targets for resonant interactions. This hypothetical approach was based on the vibrations of sound in membrane thickness. This connecting acoustics to RFR wave interactions led to Froelich's predictions of resonances at approximately 100 GHz. This is in the upper range of millimeter waves (approximately 3 mm). To date, deoxyribonucleic acid (DNA) interactions that affect the transcription and translation of functional proteins of various types are attributed to resonant DNA interactions with millimeter waves. This wavelength range can only penetrate a millimeter or less into the skin and thus presents a minimal risk to health and safety. This latter reasoning allows a well-established DNA-damaging electromagnetic radiation ultraviolet C (UVC) at 220 nm to be considered safe for disinfection because it cannot penetrate beyond the keratinized layers of the skin, based on reported effects [3].

This nonthermal mechanism assumption persists because of bioeffects in artificial preparations [4, 5]. The latter citation claims these DNA millimeter wave effects to be nonthermal but still acknowledges that experimental measurements and computational modeling showed there was "negligible" heating (+1.2 C) for the exposure time in their experimental preparations.

Induced changes in the enzymatic rates of a variety of biological pathways were claimed as millimeter wave frequency resonances, over 40 years ago, by Webb and Booth [6], Devyatkov [7], and Grundier (1977) [8]. The 1974 Russian studies that supported these conclusions were always under suspicion as propaganda to prevent the deployment of Western military communication and/or radar systems based on raising health and safety concerns. However, Gandhi et al., in 1979 [9], collected broad-frequency continuous dielectric spectroscopy measurements at millimeter-wave frequencies that showed no detectable resonances. Additionally, looking at a variety of cell functions, no effects of millimeter-wave radiation were observed that could not be attributed to preparation heating. The resonance phenomena, if they exist, as reported by Grundler et al. and as postulated by Froehlich, may represent minor cell-level bioeffects that are lost in the observed bulk dielectric data.

When I first entered the field of nonionizing electromagnetic radiation bioeffects in 1981, it was dominated by physicists and electrical engineers, on the one hand, and biologists, neuroscientists, epidemiologists, and psychologists (behavioralists), on the other hand. There were no chemists to be found. Molecular resonance at the quantum mechanical level was not even mentioned. These resonances were discussed in the realm of natural molecular vibrations in the classical sense.

Although all electromagnetic radiation can be thought of in quantum terms, the energy of RFR is so low that most of its interactions can be explained classically. The effects can be explained as wave interactions, equivalent to vibrations. Therefore, the natural frequencies of these molecules were expected to interact with the applied RFR near those same frequencies or modes.

Two of Einstein's theoretical approaches, which are not commonly known but which figured heavily in my first approach to solving the problem of designing a molecular dosimeter for RFR and microwave radiation, were his calculations on diffusion and heat capacity (specific heats of metals). The diffusion work led to his proving the physical existence of molecules, so they were no longer just a theoretical tool in chemistry. The heat capacity work led to the quantization of heat capacity in solids [10].

Einstein showed that the mean square of the distance traveled by a very small particle, such as a molecule, in random paths during a given increment of time is equal to two times the diffusion coefficient times the interval of time.

The Stokes-Einstein equation (Eq. 4.1) is used to relate the diffusion coefficient D to the radius of the particle.

Equation 4.1: Stokes-Einstein Equation

$$D = \frac{k_b T}{6\pi\eta_w r}$$

k_b = Boltzmann constant
T = temperature in degrees Kelvin
η_w = the viscosity of water at 25 °C
r = the radius of the particle

Substituting the expression of the diffusion coefficient from the Stokes-Einstein equation into Einstein's random diffusion equation, the time interval for the molecule to move the distance of its own radius can be determined (Eq. 4.2).

Equation 4.2: Stokes-Einstein Random Walk Diffusion Combined Equations

$$\Delta t = \frac{3\pi\eta_w r^3}{k_b T}$$

The cube of this element can be used to determine the volume randomly traveled through by one molecule in a given time (Eq. 4.3).

Equation 4.3: Volume Traveled Through by One Molecule in a Given Time

$$\frac{\Delta V}{\Delta t} = \frac{k_b T}{3\eta_w} \text{cm}^3 \text{s}^{-1}$$

To determine the volume swept out by a mole of particles in a given time (or half time), the molar gas constant is substituted for the Boltzmann constant. This equation is modified to show the effect of the volume between large minimally diffusible particles interacting with much smaller (molecular or atomic-sized particles) reactive diffusible particles of a half-life limited to the distance traveled to the larger particles' surface. I substituted the difference between the suspension volume V and the large non-diffusible particles' volume, V_c, or the volume element in the previous equation. The factor c, the number of moles of small reactive diffusible particles in $V-V_c$, must also be included in this final equation (EQ. 4.4).

Equation 4.4: Half-Life of Diffusion of a Small Molecule

$$t_{\frac{1}{2}} = \frac{V - V_c}{c\left(RT/3\eta_w\right)}$$

$$s = \frac{cm^3}{moles\left(\frac{cm^2\, g\, s^{-2}\, mole^{-1}\, K^{-1}\, K}{g\, cm^{-1}\, s^{-1}}\right)}$$

From these calculations, one can see that those large biological molecules, or those fixed in membranes or superstructures, simply have vanishingly small diffusion coefficients and exceedingly long diffusion times. Any reactions that require molecular collisions or substantial kinetic energy to occur must involve small molecules impinging on these large biological molecules and structures or within such molecules or structures.

Any other form of interaction beyond rotational would require the RFR fields to drive small molecules into these structures or captive ionic components through the biological polymers or structures.

These conclusions and assumptions led to the next consideration of Einstein's calculations of quantization of heat capacity in a solid to explain specific biological macromolecule interactions with RFR. This latter theory connected the natural vibration frequency of a macrostructure, a crystal, to a quantum effect, the absorption of quasiparticles of energy called phonons, like photons. The RFR photon energy is first converted into vibrational energy, and then the energy is absorbed in discrete packets of vibrational energy called phonons (ε), to which quantum calculations can be applied to affect the macroproperty, heat capacity (Eqs. 4.5, 4.6, and 4.7). This theoretical approach was the first attempt to link the classical electromagnetic effects of RFR to quantum theory. However, it applies to solids, not liquids. If it applied to DALM, then DALM must be acting as a solid in a liquid, not a solute in a solution.

Equation 4.5: Heat Capacity in a Constant Volume

$$\mathbf{C_V = 3Nk\left(\varepsilon / 2kT\right)}$$

Equation 4.6: Quantization of RFR Absorbed Energy

$$\mathbf{C_V = \left[\left\{3Nk\left(\varepsilon / 2kT\right)^2\right\}\left\{e^{\varepsilon/kT}\right\}\right] / \left(e^{\varepsilon/kT} - 1\right)^2\}}$$

As T → ∞
Then

Equation 4.7: Heat Capacity in a Solid Crystal

$$C_V = 3Nk$$

On a molar basis:

$$C_V = 3Nk = 3nR$$

At the high-temperature limit, when T > > θ_E, Einstein's heat capacity reduces to Cv = 3Nk (the Dulong and Petit law).

At the low-temperature limit, when T < < θ_E, C_v goes to zero as T goes to zero, as determined by the third law of thermodynamics.

C_V = heat capacity at a constant volume
ε = quantum energy of a phonon = ħω
ω = phonon vibrational frequency
k or k_b = Boltzmann's constant
N = number of atoms with 3 N motions in a crystal matrix
R = molar gas constant
n = number of moles
θ_E = Einstein's temperature

The Petit and Dulong classical explanation (empirical model), when compared to the calculations and graphics of Einstein's and Debye's theories, as well as discrete vibrational models, merges with them within certain temperature ranges [11, 12]. Einstein's theory leads to quantum mechanical effects but merges with the classical data at higher temperatures. The Einstein and Debye models employ temperature scales that correlate with the vibrational states of the atoms or molecules in the crystal lattice.

In 1912, Debye realized that, unlike Einstein's model, low-energetic excitations of solid material were not oscillations of a single atom but rather collective modes propagating through the material. These vibrations are equivalent to sound waves, with their propagation speed being the speed of sound in the respective material. The modes, showing quantum behavior, only accept energy in discrete amounts. We will see later in this chapter that the data support this concept of producing sound waves in spite of the fact that the DALM is in a liquid environment. Later, we will see that this is tied to the unique properties of water (Chap. 7).

Einstein made assumptions about the heat capacity of solids:

1. Each solid is composed of a lattice of N atoms.
2. Each atom is moving independently in three dimensions within the lattice (with 3 degrees of freedom).

3. The atoms inside the solid lattice do not interact with each other.
4. All the atoms inside the solid vibrate at the same frequency.

The entire lattice's vibrational motion has a total of 3 N motions or degrees of freedom. The difference between Einstein's and Debye's models was based on the single, independent atom vibrations versus a collective, coordinated vibration of atoms (sound transmission). When the natural frequencies of biological molecules were being determined to predict RFR frequencies that would interact with them, acoustics were used that could fit either of these models.

This approach fits the Debye model more than the Einstein model. However, the forms of DALM and the polymeric structure containing individual sites of phonon absorption may lead to a combination of both models. The origin of sound may be in individual polymeric fibers.

Einstein's model was accurate, in part, because the experimental data supported his hypothesis. However, his assumption that atoms inside a solid could not interact did not allow sound to propagate through it. The mathematical definitions of these two models' temperature scales are shown in Eqs. 4.8 and 4.9, respectively.

Equation 4.8: Einstein Temperature (θ_ε)

$$\theta_\varepsilon = \mathbf{h}\nu / \mathbf{k}$$

$$\theta_\varepsilon / \mathbf{T} >> \mathbf{1}$$

$$\mathbf{C_V} = \mathbf{3Nk}\ (\theta_\varepsilon / \mathbf{T})^2\ \mathbf{e}^{-\theta\varepsilon/\mathrm{T}}$$

$$\nu = (\mathbf{1}/\mathbf{2}\pi)\sqrt{}\,\mathbf{k}/\mu$$

Equation 4.9: Debye Temperature

$$\mathbf{T_D} = \hbar\omega_\mathbf{p} / \mathbf{k_b} = \hbar\omega_\mathbf{D} / \mathbf{k_B} =$$

$$(\hbar\nu_\mathbf{s} / \mathbf{k_B})(\mathbf{6}\pi^2\mathbf{N}/\mathbf{V})^{1/3}$$

ν_S = velocity of sound
$\omega_D = \nu_S\ (6\pi^2 N/V)^{1/3}$
n_E = number of phonons
$n_E = 1/(e^{E/kbT} - 1) = 1/(e^{\hbar\omega/T} - 1)$
ν = oscillator frequency of atoms inside a solid
V = volume

Debye heat capacity (constant volume)

$$\mathbf{C_V = \left(12\pi^4 Nk_b T^3\right)/5T_D{}^3}$$

They both merge with the Petit-Dulong empirical model at higher temperatures. The absorption of specific vibrational frequencies into a solid can be enhanced even more with finer divided discrete energy levels of rotation. This consequence of the previous discussions will become very important later in explaining the behaviors of DALM.

The absorption of phonons would require a crystalline structure with vibrational levels that can absorb only certain discrete integer multiples of lower levels. The rigid ribbon, triclinic quasicrystalline structure of DALM may meet this criterion. Additionally, the quantization of heat capacity has required temperatures very much below kT. At higher temperatures, the equipartition of energy comes into effect. This theorem states that particles of the same energy are of equal probability, whether they have vibrational, translational, or rotational degrees of freedom.

The structure of DALM has fewer degrees of freedom, acting like molecules at much lower temperatures because of its long linear, planar nature with energy-barrier-restricted rotations (intramolecular steric hindrance) [13]. If phonon absorption is responsible for the steady-state luminescence of DALM, then N is equal to the number of photons or heat-equivalent phonons; k, also k_b, is Boltzmann's constant; k_o is the photon-per-second rate (zero-order rate constant of steady-state light emission at a given temperature); SAR is the specific absorption rate in watts (joules per second); 6 is conductivity; and E is the electric field. This first quantum mechanical pass at a mechanism computationally connects with classical SAR calculations (Eqs. 4.10 and 4.11).

Equation 4.10: Einstein's Quantization of Heat Capacity and Its Relationship to SAR

In 1907 Einstein, in the first application of quantum theory to ta problem other than radiation, modeled a solid body containing N atoms as a collection of 3 N harmonic oscillators.

$$C_V(T) = \frac{dU}{dT} = -k_B\beta^2\frac{\partial U}{\partial\beta} = [U = 3NE] = 3Nk_B(\beta h\nu)^2\frac{\exp(\beta h\nu)}{[\exp(\beta h\nu)-1]^2}$$

$$\text{high}T(k_BT >> h\nu):\quad C_V(T) = 3Nk_B \quad \text{equipartion}$$

$$\text{low}T(k_BT << h\nu):\quad C_V(T) = 3Nk_B(\beta h\nu)^2\exp(-\beta h\nu) \propto \exp(-T)$$

or

$$c = 5.94 \sum_i \frac{\left(hv_i / kT\right)^2 e^{hv/kT}}{\left(e^{hv_i/kT} - 1\right)^2}$$

$$\begin{aligned} Nhv &= 2Nk\left(\ln k_o\right)^2 \\ SAR &= \sigma E^2 \\ \therefore \sigma &= 2N; E = \ln k_o \end{aligned}$$

Equation 4.11: SAR Relationship to Volume, E Field and Conductivity

$$\mathrm{SAR} = \frac{1}{V} \int_{\text{sample}} \frac{\sigma(\mathrm{r})\left|\mathrm{E}(\mathrm{r})\right|^2}{\rho(\mathrm{r})} d\mathrm{r}$$

Where

σ is the sample electrical conductivity
E is the (RMS = root mean square) electric field
ρ is the sample density
V is the volume of the sample

The steady-state luminescence strongly correlates with temperature and energy absorption in agreement with this mechanistic model, but as we will see, it is too limited to explain all the RFR interaction responses of DALM.

A further examination of the characteristics of DALM responses, RFR, and thermal induction suggested that there needed to be a more basic mechanistic explanation.

The experimental phenomena illustrate the departure of DALM's behavior from classical, requiring a more inclusive broader theoretical explanation [13].

Fig. 4.1 is a series of video frames of the pulsed microwave exposure of a DALM peroxidizing solution. The different intensities of the luminescence in the frames indicate the periodicity of the pulsed microwaves.

In Figs. 4.2 and 4.3, tracings show the correlation of sound and light pulses with microwave pulses, which are building in intensity and declining as such after a major flash. Even pulsed direct current applied to the DALM solutions demonstrated the correlation of luminescence with the electric field (Figs. 4.4 and 4.5).

The DALM immobilized on paper or an epoxy plastic polymer fluoresced when exposed to UVA (315–400 nm range), but slowly if it was exposed to UV and heated simultaneously, then cooled, and re-exposed to the same UV. It fluoresced after the

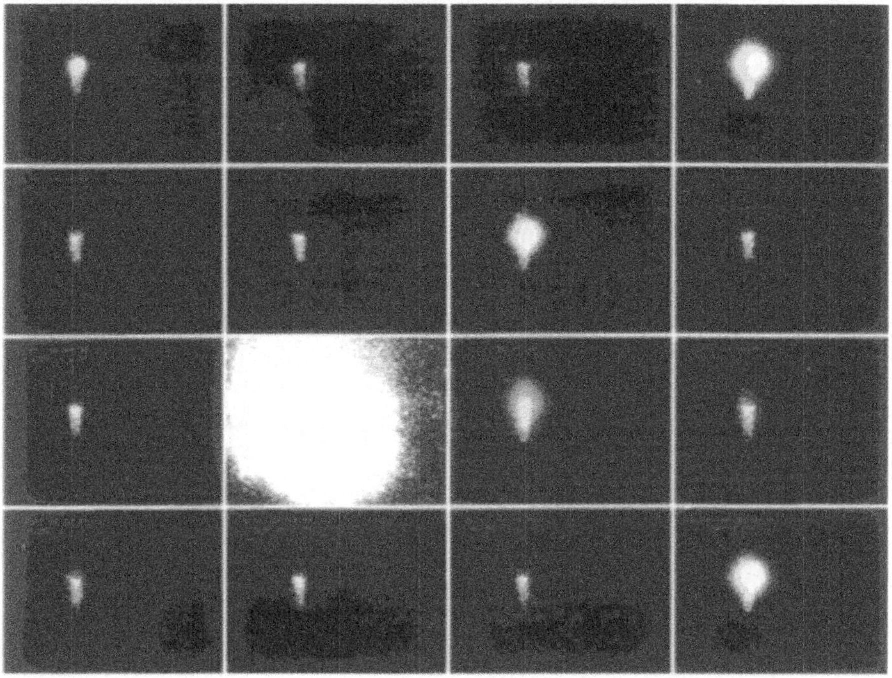

Fig. 4.1 A series of video frames of the pulsed microwave luminescence responses of an activated DALM solution showing the building and recovery of the intensity (periodicity)

light was turned off for periods of time that positively correlated with the length of time of simultaneous heating and UV exposure. This process is already known in other materials and is referred to as thermally activated delayed fluorescence (TADF) (Fig. 4.6) [14, 15].

Finally, when measuring the killing of bacterial spores in a solution by peroxidizing DALM in the aqueous solution, we found, counterintuitively, that the solution had to be cooled below 90 °C, decreasing luminescence for the killing to be maximized. These responses clearly showed that the effects of RFR and microwaves on DALM were not purely thermal. What new theory could account for this?

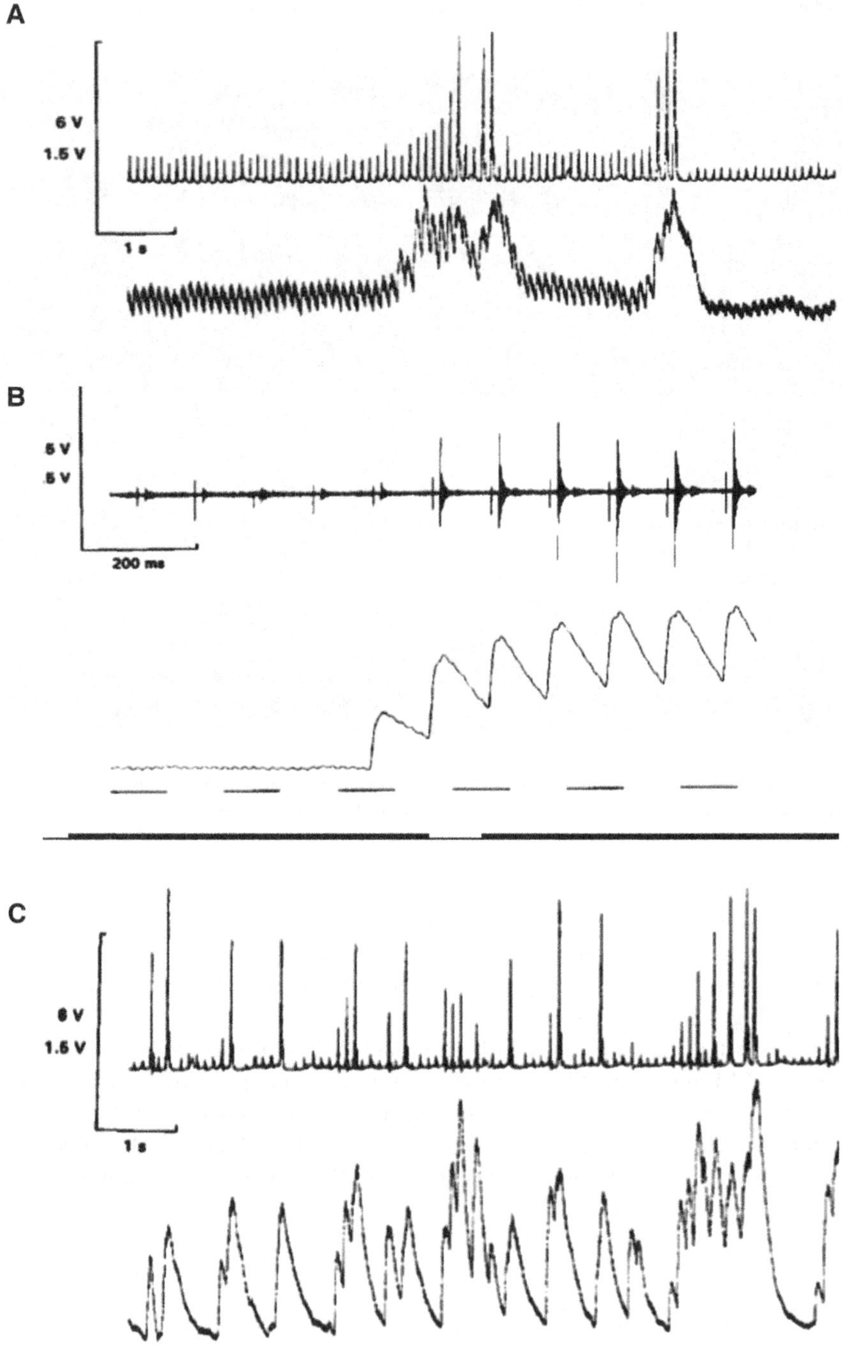

Fig. 4.2 (**a**) Sound and correlated light pulses. The regularity of the pulses synchronous with the microwave pulses indicated that the response was not a simple sound but was produced directly by the RFR pulses. The large plasma discharges are superimposed on the smaller ones that correlate with the microwave, sound, and light pulses. (**b**) The sound and light pulses of a luminol, hydrogen peroxide, and sodium carbonate mixture without DALM, showing chaotic sound and light pulses, which are indicative of boiling and/or gas release

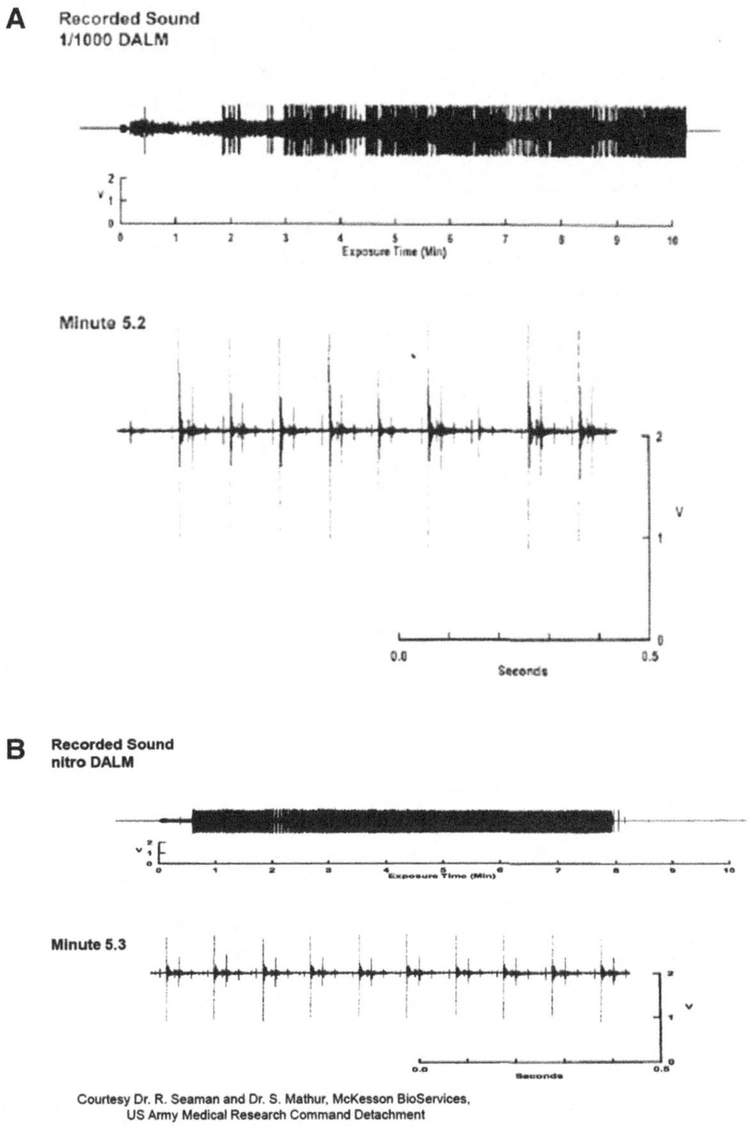

Fig. 4.3 (**a**) and (**b**) Activated DALM sound pulses (A, first tracing) and the effect of chemically modifying the DALM polymer by additional nitration (B, second tracing; adding nitro groups to the aromatic rings). The pulse responses of light and sound are even more uniform and consistent

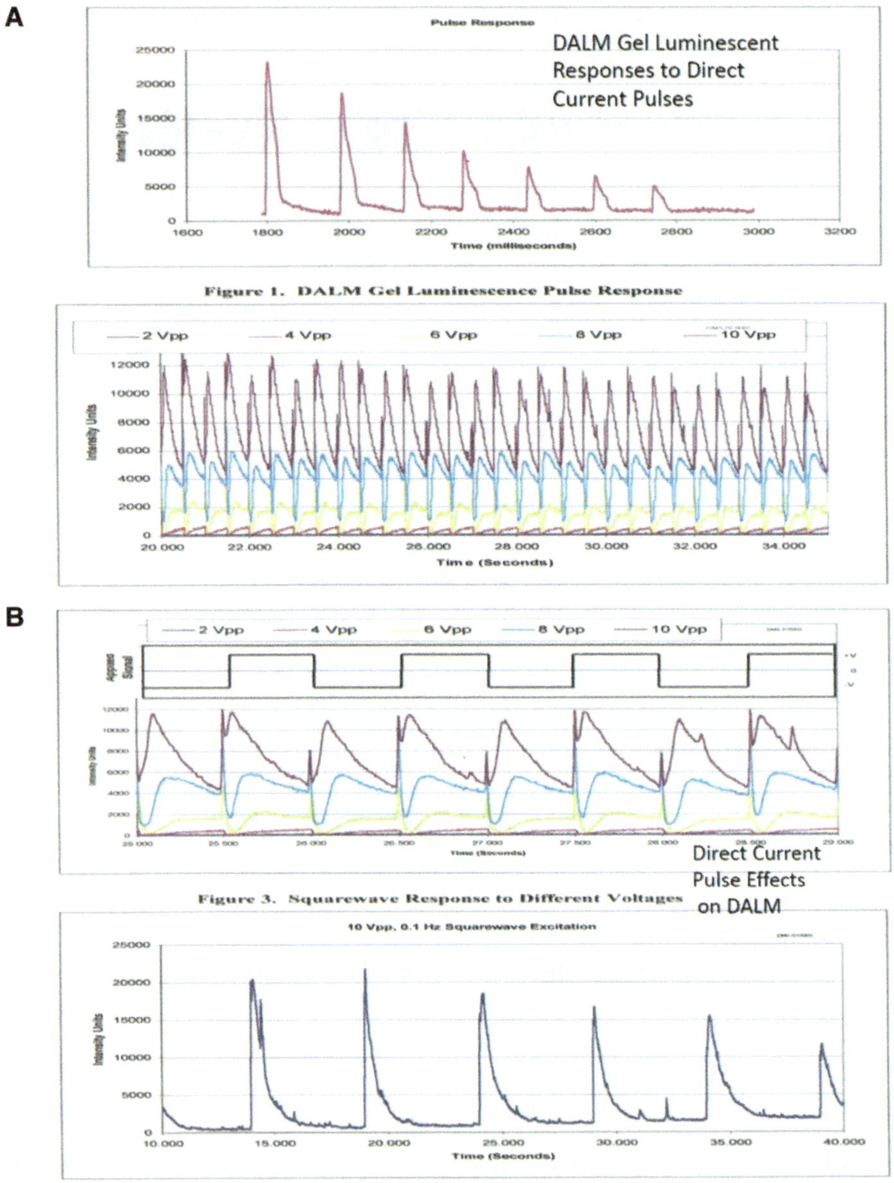

Fig. 4.4 (**a**) and (**b**) Direct current (DC) voltage when pulsed as square waves created light pulses that decreased in amplitude if they were timed too closely, but the baseline luminescence rose and remained continuous. The luminescence was voltage dependent and increased and fell in a periodic manner when the pulses at a given voltage were close together

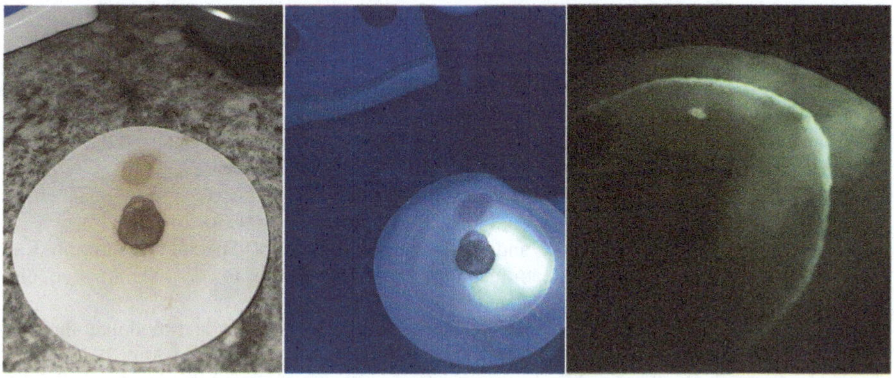

A B C

Fig. 4.5 (**a**), (**b**), and (**c**) The first photograph shows a DALM liquid preparation in visible light that has laterally diffused from a concentrated application to filter paper; the second photograph shows DALM under a UV flashlight, displaying immediate blue fluorescence; the last photograph shows thermally activated delayed green fluorescence (TADF) of the less concentrated (and probably lower molecular mass components of the polymer mixture) DALM, removing autoquenching

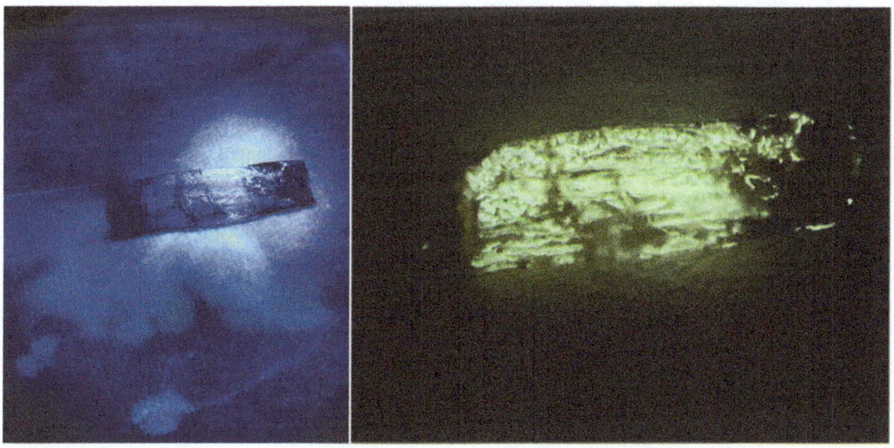

A B

Fig. 4.6 (**a**) and (**b**) The first photograph shows blue fluorescence of DALM and iron micro- and nanoparticles bound to epoxy on a polystyrene substrate; the second photograph is of yellow-green TADF of the same material

References

1. Froehlich, H., Collective behavior of nonlinearly coupled oscillating fields. J Collective Phen 1:101–109, 1973.
2. Froehlich, H., The extraordinary dielectric properties of biological materials and the action of enzymes. Proc Natl Acad Sci USA 72:4211–4215, 1975.
3. Freeman, S., et al., Systematic evaluating and modeling of SARS-CoV-2 UVC disinfection, Scientific Reports 12: 5869, https://doi.org/10.1038/s41598-022-09930, 2022.
4. Panagopoulos D., Johansson O., and Carlo G., Evaluation of Specific Absorption Rate as a Dosimetric Quantity for Electromagnetic Fields Bioeffects. PLoS ONE 8(6): e62663. doi:https://doi.org/10.1371/journal.pone.0062663, 2013.
5. Lawler, N., et al., Millimeter waves alter DNA secondary structures and modulate the transcriptome in human fibroblasts, Biomedical Optics Express: 13 (5),3131, 1 May 2022.
6. Webb, S., and Booth, A., Microwave absorption by normal and tumor cells, Science 174:72–74, 1971.
7. Devyatkov, N., influence of millimeter-band electromagnetic radiation on biological objects (transl.), Sov Phys USPEKHI 16:568–579, 1974.
8. Grundler, W., et al., Resonant growth rate response of yeast cells irradiated by weak microwaves, Phys. Lett 62A: 463–466, 1977.
9. Gandhi, O., et al. Millimeter wave and Raman spectra of living cells—some problems and results. Workshop on Mechanisms of Microwave Biological Effects, University of Maryland, College Park, MD, 14-16 May 1979.
10. Kiel, J., The Cytotoxic Activity of Peroxidases, PhD Dissertation, Texas Tech Health Sciences Center, School of Medicine. University Microfilms,1981.
11. 17.4: Heat Capacity at Constant Volume is the Change in Internal Energy with Temperature; https://batch.libretexts.org/print/url=https://chem.libretexts.org/Bookshelves/Physical_and_Theoretical_Chemistry_Textbook_Maps/Physical_Chemistry_(LibreTexts)/17%3A_Boltzmann_Factor_and_Partition_Functions/17.04%3A_Heat_Capacity_at_Constant_Volume.pdf.
12. Debye Model for Specific Heat—Engineering LibreTexts, https://batch.libretexts.org/print/url=https://eng.libretexts.org/Bookshelves/Materials_Science/Supplemental_Modules_(Materials_Science)/Electronic_Properties/Debye_Model_For_Specific_Heat.pdf
13. Brown, K., Ferris, K., Irving, G., COMPUTATIONAL CHEMISTRY FOR THE HIGH POWER MICROWAVE INITIATIVE, AFRL-HE-BR-TR-1999-0224, October 1999.
14. TADF: What is Thermally Activated Delayed Fluorescence?: Scientific Article, https://www.edinst.com/blog/tadf-thermally-activated-delayed-fluorescence/
15. Yin, X., et al., Recent Advances in Thermally Activated Delayed Fluorescent Polymer—Molecular Designing Strategies, Frontiers in Chemistry, www.frontiersin.org,, Volume 8, Article 725, August 2020.

Chapter 5
Splitting Hairs: Fine and Hyperfine Splitting and Free Radicals

Abstract Are the properties of the polymer diazoluminomelanin unique, or do they illustrate fundamental principles of radiofrequency radiation (RFR) interaction with biological and other macromolecules? Is it a coincidence that diazoluminomelanin (DALM) can be biosynthesized by nitrate reductase and/or nitric oxide synthase and flavoprotein participation, or are there no "coincidences?" Nitric oxide synthase (NOS) has shown some nonthermal sensitivity to ultrawide band nonionizing electromagnetic radiation (NEMR). Nitric oxide itself is a stable free radical gas that can have nonthermal quantal interaction with NEMR. One explanation may be found in the flavin in NOS and its unique biochemistry. We will examine a key quantum mechanical candidate for this interaction and how it relates to both the enzymes and DALM. This chapter is a transition from the fundamental physical science discussion in previous chapters (representing the physicists, electrical engineers, and chemists involved in bioelectromagnetics) to the biochemistry and molecular biology discussions and bionanotechnology applications in the second half of the book. Both sides (physical scientists versus molecular biologists) must be patient with and understand at least the basics of each other's science to grasp the full story of radio-controlled bionanotechnology.

Keywords Flavin adenine dinucleotide (FAD) · Nitrate reductase · Rotations · Vibrations · Jablonski diagram · Intersystem crossing · States · Lossless · Free radicals · Hyperfine splitting · Cryptochrome

The search for a molecular mode resonator for radiofrequency radiation (RFR) and microwave energy began for me in the United States Air Force (USAF) in 1981 and ended in 2011. Today, little has happened to change the bioelectromagnetics research community's search for this "holy grail" of RFR bioeffects. In 2011, I participated in a workshop at the Electric Power Research Institute titled "Program on Technology Innovation: Health Issues Related to Radiofrequency Emissions from Smart Technologies Workshop."

Asher Sheppard and I led the workshop discussion on mechanistic studies of RFR bioeffects. We examined physical, biophysical, and chemical theory and data

J. L. Kiel, *Bionanotechnology*, https://doi.org/10.1007/978-3-031-63006-4_5

from in vitro experiments, tissue culture, and laboratory animals. We covered much of what has been discussed in earlier chapters and also, what is of interest here, the possible nonthermal interactions with specific molecules that overcome the limitations of very low energy absorption of bulk tissues, much below kT. Barring specialized organs containing specialized thermosensitive cells, such as the heat-sensing facial pits of pit vipers and pythons, which can respond to temperature differences less than 0.1 K and others as little as 0.003 K, there are a few, if any, mechanisms that can explain bioeffects at these energies. Temperature becomes irrelevant at cellular to molecular dimensions because of rapid thermal diffusion for differences as little as 10^{-6} K, eliminating temperature as anything but a bulk effect on a population of molecules or the total of many cell organelles. The major conclusion was that no physical mechanism exists for low-level, nonthermal exposures between 10 MHz and hundreds of GHz, except for free-radical spin-state interconversion.

We concluded that three mechanisms represent the possibilities that should be investigated: (1) heating (the most historically investigated), (2) enhanced RFR energy absorption, and (3) dynamic changes involving charged particles and biochemical processes, including ion transport, metabolic energy generation, and neurotransmission (enzyme kinetic effects) [1]. The specific list of these mechanisms is given in the workshop summary report. The research priorities arrived at by consensus at the workshop were (1) pulsed versus continuous wave effects on the electroencephalogram (EEG), particularly alpha rhythms; (2) small thermal effects because of the peculiar heating patterns of RFR and possible responses to temperatures less than 1 °C; and (3) effects on cryptochromes (which contain the cofactor flavin adenine dinucleotide (FAD), important to our discussion on the biological consequences of the intersystem crossing of spin states). Cryptochromes are a class of blue-light-sensitive flavoproteins that are key to controlling circadian rhythms in plants and animals and sensing the earth's magnetic field for birds and sea turtles in migration (navigation) [2]. The one to be discussed here and the most relevant to the final application is the intersystem crossing of spin states in redox molecules, including enzymes and their cofactors, especially FAD (Fig. 5.1) and protoporphyrin IX hemes (Figs. 5.2 and 5.3).

FAD is central in a variety of energy-generating and -capturing electron transport systems and enzymes [2]. These redox activities include six mechanisms, two of which form transient free radicals: (1) hydride transfer through the separate transfer of two electrons and a hydrogen ion (proton), as in mitochondria; (2) hydride transfer as a negative ion from nicotinamide adenine dinucleotide hydride (NADH) or nicotinamide adenine dinucleotide phosphate (NADPH) in dehydrogenases and reductases; (3) radical formation by electron abstraction, forming a stable tertiary carbon-centered free radical; (4) loss of hydride to an electron-withdrawing organic group; (5) nucleophilic addition to break a carbon-carbon bond; and (6) formation of a carbon radical on the flavin that reacts with oxygen and acid to form hydrogen peroxide, as in glucose oxidase. Examples of enzymes that use FAD are nitric oxide synthase, which makes nitric oxide for neurotransmission, transduction for insulin release, antimicrobial activity, and vasodilation; NADPH oxidase, which produces superoxide as an antimicrobial in granulocytes; monoamine oxidase (MAO), which

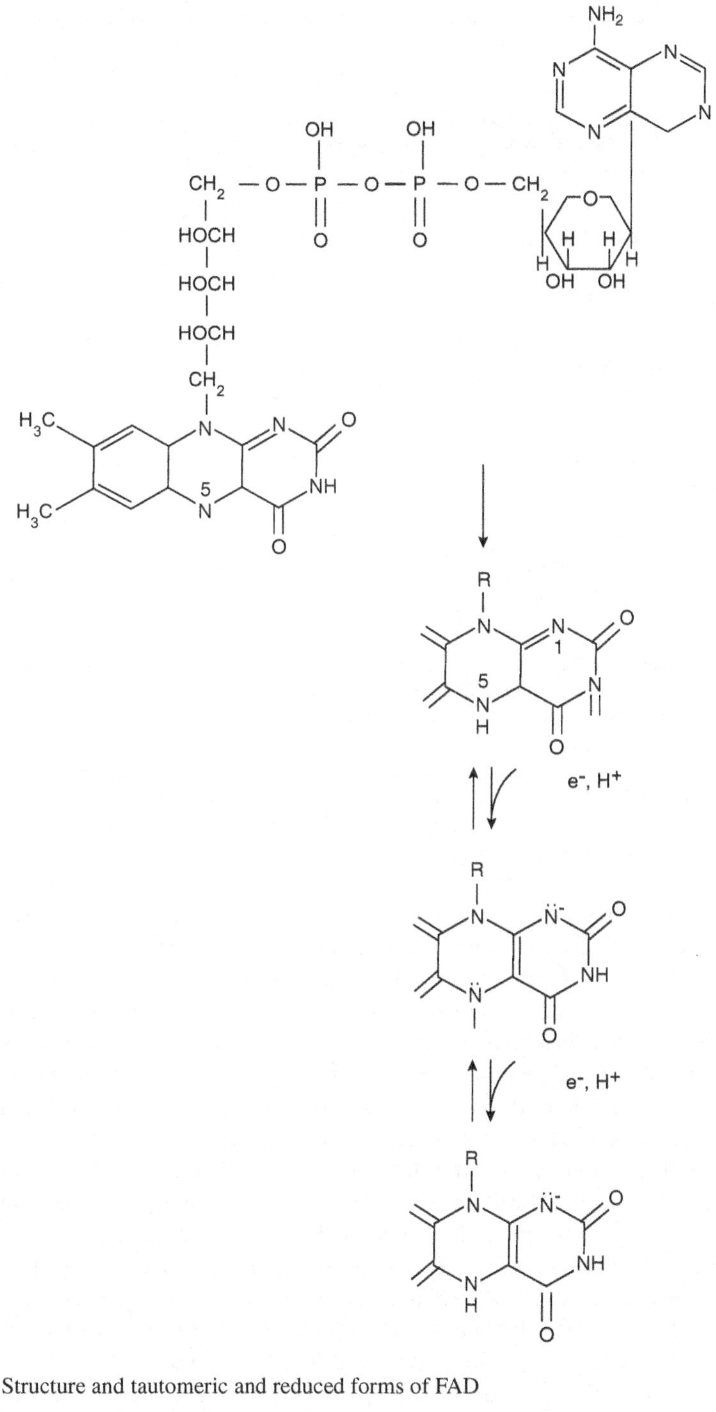

Fig. 5.1 Structure and tautomeric and reduced forms of FAD

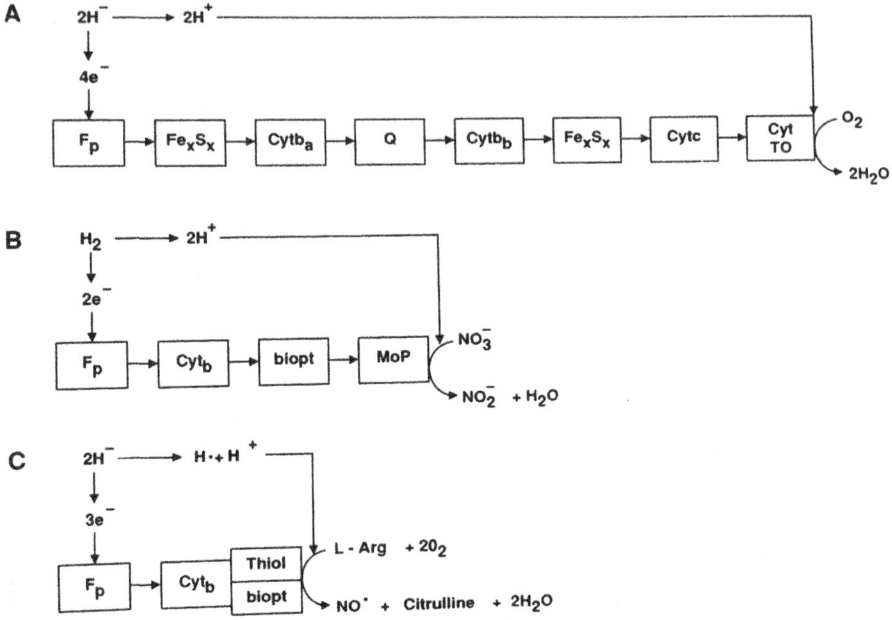

Fig. 5.2 Comparative diagrams illustrate the close structural and basic functional properties of three FAD/cytochrome b systems with otherwise widely diverse physiological functions. (**a**) Mitochondrial; (**b**) nitrate reductase; and (**c**) nitric oxide synthase. *Fp* flavoprotein, *Cytb* various type-b cytochromes, *FexSx* nonheme-iron-sulfur proteins, *Q* ubiquinone, *Cytc* cytochrome c, *Cyt TO* terminal cytochrome oxidase, *biopt* biopterin, *MOP* molybdenum-carrying cofactor, *L-Arg* L-arginine. (From Type B Cytochromes: Sensors and Switches, CRC Press, Taylor & Francis Group, 1995 and 2018; illustration by Ms. Arlene M. Schirmer (AL/DOSP), USAF Armstrong Laboratory)

catabolizes norepinephrine, serotonin, and dopamine; glucose oxidase, which oxidizes glucose to D-glucono-delta-lactone and hydrogen peroxide; prenylcysteine lyase, which cleaves prenylcysteine (a posttranslational protein modification) to yield isoprenoid aldehyde and cysteine (residue in protein); glutathione reductase, which converts glutathione disulfide into glutathione as a subsequent substrate for antioxidant and detoxifying cellular activities; cofactors for cytochrome P450s for hydroxylations and detoxification (metabolism of many drugs); and many other chemicals. FAD is in 84% of 90 human flavoproteins.

Oxidized FAD is aromatic, and the reduced form is not aromatic. The unbound form is fluorescent, but the fluorescence is quenched upon protein binding. Fully oxidized FAD is yellow, and half reduced is blue to red depending on pH, with the fully reduced (high-energy) form being colorless. The fully oxidized state is vulnerable to nucleophilic attack, the reduced form is polarizable, and the half-reduced form is unstable in water. FAD has a major optical absorption at 450 nm. Oxidized FAD absorbs at 450 nm and fluoresces from 515 to 520 nm.

Cryptochromes form a special group of FAD proteins, some of which have photo-driven deoxyribonucleic acid (DNA) repair enzymatic activity (photolyases)

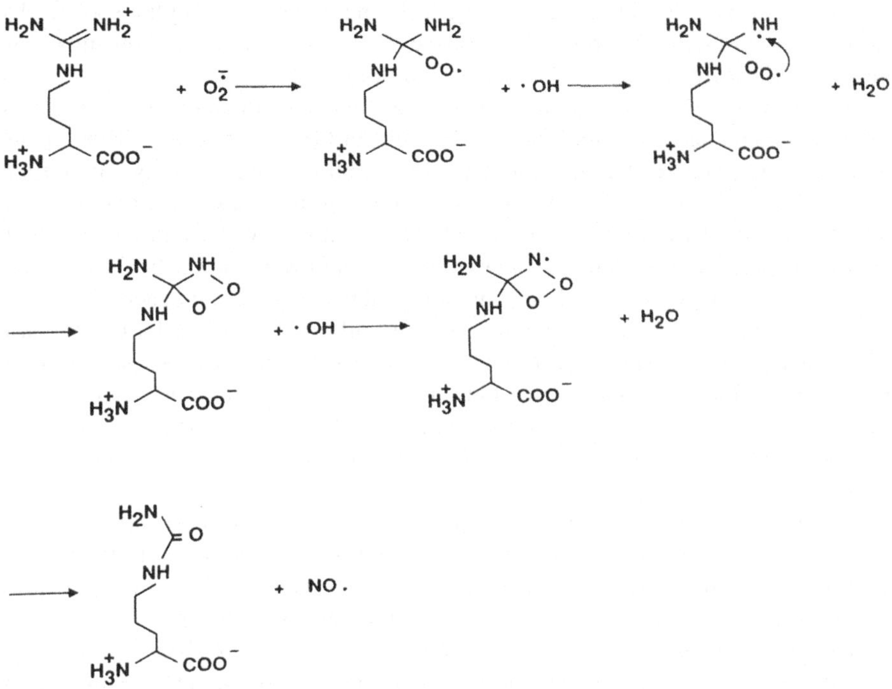

Fig. 5.3 The complex 5 electron oxidation of arginine to produce nitric oxide by the synthase and the net reaction of nitric oxide synthase with oxygen. (From Type B Cytochromes: Sensors and Switches, CRC Press, Taylor & Francis Group, 1995 and 2018. CRC Press; illustration by Ms. Arlene M. Schirmer (/DOSP), USAF Armstrong Laboratory)

and others that act as transactivating factors or both [3–5]. They, as mentioned earlier, are an important potential biological molecular target for specific electromagnetic field sensitivity found in a wide variety of organisms, including bacteria, plants, animals, and humans.

The cryptochrome/photolyase group of proteins transactivates or transduces adaptive responses to ultraviolet (UV) and blue-light exposure across all kingdoms of living things [3]. The photolyase moiety in these proteins may be conserved but nonfunctional (especially in humans), but when functional, it is predominantly active in the DNA repair of cyclobutane pyrimidine dimers and 6–4 photolesions caused by UV radiation. Cryptochromes transduce signals contributing to broad functions, including growth, development, magnetosensitivity, and circadian rhythms. Either one of two possible photoactive cofactors is found in the one of two main classes of photolyases, either pterin, methenyltetrahydrofolate, in folate photolyases or deazaflavin, 8-hydroxy 7,8-didemethyl-5-deazariboflavin, in deazaflavin photolyases. FAD is the only cofactor required for enzymatic activity. However, these secondary cofactors greatly increase reaction rates in low light. FAD and other cofactors have basic structural and chemical activities similar to those of DALM, which may demonstrate the broad application to RFR bioeffects represented by this otherwise unique polymer.

We must revisit the fundamentals one last time before entering the specifically applied world of radio-controlled vectors; we must be made aware of how important the quantum world and fine and hyperfine splitting are to that goal.

It is this book's intention not to be an instructive text on quantum theory but only to relate it to the application of the radio control of biological vectors. Three quantum properties then need to be examined: (1) the splitting of the energy levels (transitioned by microwave energy) of triplet states by spin and electron angular momenta interactions and hyperfine splitting into more quantum levels by interaction with atomic nuclear spins enhanced by externally applied magnetic fields (Zeeman effect, the splitting of spectral lines into components by a static magnetic field); (2) "lossless" singlet to triplet and the reverse, "forbidden" transitions based on phase (time) changes in respective wave functions; and (3) mixed wave functions leading to "unobservable" entangled states biased by magnetic fields. The connection of these to the kinetics of redox reactions and photochemistry of DALM will be discussed. Singlet refers to a pair of electrons with opposite spins, and triplet refers to a pair with the same spin.

That there is a link is strongly suggested by the departure of DALM photochemistry from classic thermochemiluminescence and the discontinuity of effects, suggesting quantal state transitions. The fluorescence is enhanced by previous heating or microwave exposure in conjunction with simultaneous exposure to ultraviolet A (UVA: 315–400 nm wavelengths) light and rapid cooling thereafter. Also, the 100- to 1000-fold increase in bacterial killing by exposure to activated DALM and microwaves while cooling, as opposed to allowing heating, is counterintuitive to classical theory. These observations cannot be explained by classic electromagnetic theory alone. Finally, predictive wave function theory supports vibrational and rotational states quantized, coupled, and mixed with the electronic states in the observed NEMR responses. These latter states connect to my phonon discussions in Chap. 4 and the possibility that the conversion of RFR and microwaves into heat can still allow them to be absorbed as quanta.

A basic tenet of quantum mechanics is that only certain energy levels and locations of electrons can exist as well as the transition of these to higher levels. The wave functions can only be defined for the simplest atoms such as hydrogen and approximated for complex organic macromolecules such as DALM (using the Hartree-Fock method for the approximation of a wave function and energy of a quantum many-body system in a stationary state).

This problem necessitates that we first illustrate it with simple mathematical models from which we can draw assumptions about the more complex cases. A wave function can be envisioned as a probability function along a single axis within the confines of two insurmountable barriers, "walls." It contains a real and imaginary part, completing a continuous standing wave or one reflected off the walls of the "box." The periodicity is determined by the position and momentum of the particle traveling back and forth between the walls of the box, and the probability of the particle being present at any given position along the trajectory is determined by the square of the function.

The total probability of the particle being in the box is one, and outside it is zero. This very simple notional case represents a single dimension particle (point) that is spinless. Spin, if added to the description, can be seen as a "reflection" (symmetry) of the particle, parity.

Similar to the pairing of momentum Δp (mass times the velocity of the particle, mv) with (the location along the single dimension) Δx, the pairing of ΔE and Δt, energy and time (including lifetime τ) cannot be determined simultaneously because of the Heisenberg uncertainty principle (Eq. 5.1). This equation states the uncertainty in momentum (delta p), and location (x) is greater than (cannot be less) than the reduced Planck's constant divided by 2. The reduced Planck's constant, $\hbar$, is Planck's constant, h, divided by 2π, that is, the quantization of angular momentum. Planck's constant is equal to $6.62607015 \times 10^{-34}$ J·Hz^{-1} or $4.135667696 \times 10^{-15}$ eV·Hz^{-1}.

Equation 5.1: Heisenberg Uncertainty Principle

$$\Delta p \Delta x > \hbar / 2$$

If time is known, energy (E) cannot be determined with any precision, and when E is known precisely, time, or lifetime τ, cannot be determined. When energy is radiated from the excited state of this particle, it is in a bandwidth rather than a specific value, as determined in Eq. 5.2.

Equation 5.2: Uncertainty of Radiated Energy

$$\Delta E = \hbar / \tau$$

The frequency of radiation emitted from the particle (with the radius r), the excited-to-ground state, varies within a defined bandwidth (Eq. 5.3).

Equation 5.3: Uncertainty of Radiated Frequency

$$\Delta v = \Delta E / \hbar = (1/2)\pi r$$

Because of the uncertainty principle, the zero-point energy of a particle cannot be zero: ω = the angular velocity. The one-dimensional wave function can be extended to a radial three-dimensional one described in Eqs. 5.4, 5.5, and 5.6 (E_{min} = zero-point energy; t = time).

Equation 5.4: Zero-Point Energy

$$E_{min} = \hbar\omega / 2$$

Equation 5.5: Wavefunctions

$$\int \left|\Psi(r,t)\right| 2dr = 1$$

Equation 5.6: Wavefunction Values

$$\Psi(\pm a) = \pm A \sin(k_a) + B \cos(k_a)$$

$$k_a = n\pi / 2$$

where r = the radius of the sphere and dr determines a volume element of the sphere. The solution is the position probability density normalized to a total probability of 1 (100% of finding the particle in the volume). The wave equation is $\Psi(x) = 0$ at the walls, where $x = +$ or $-a$. Therefore, the general solutions of the wave equation are Eq. 5.6.

There are two classes of solutions: $A = 0$ and $\cos ka = 0$ and $B = 0$ and $\sin k_a = 0$, where n is an odd integer (quantum number) for the first class and n is an even integer (quantum number) for the second class. The solutions of the two classes are in Eqs. 5.7 and 5.8. These wavefunctions, representing different states of the same or close energies, are equal in value but not interchangeable or converted into one another by energy level or the frequency of electromagnetic wave changes. Therefore, the electronic energy-level changes are by discrete energy inputs (quanta), and the crossing between the even and odd states is "forbidden," except for a peculiarity of quantum mechanics.

Equation 5.7: Wavefunctions for Odd Values

$$\Psi \mathbf{n}(\mathbf{x}) = \mathbf{B} \cos(\mathbf{n}\pi\mathbf{x} / \mathbf{2a})$$

$$\mathbf{n} = \mathbf{1, 3, 5, \ldots}$$

Equation 5.8: Wavefunctions for Even Values

$$\Psi \mathbf{n}(\mathbf{x}) = \mathbf{A} \sin(\mathbf{n}\pi\mathbf{x} / \mathbf{2a})$$

$$\mathbf{n} = \mathbf{2, 4, 6, \ldots}$$

For the odd and even integers (quantum levels), if we make their probabilities 1, Eqs. 5.9 and 5.10 would apply.

Equation 5.9: Wavefunctions for Odd Values with Probability of 1

$$\Psi \mathbf{n}(\mathbf{x}) = \mathbf{1} / \mathbf{a}^{1/2} \cos(\mathbf{n}\pi\mathbf{x} / \mathbf{2a})$$

$$\mathbf{n} = \mathbf{1, 3, 5, \ldots}$$

Equation 5.10: Wavefunctions for Even Values with Probability of 1

$$\mathbf{\Psi n(x) = 1 / a^{1/2} \sin(n\pi x / 2a)}$$

$$\mathbf{n = 2, 4, 6, \ldots}$$

These two classes exist independently but not simultaneously. Inputting more energy only raises the level to the next even or odd level but does not lead to a crossover between odd and even solutions (forbidden). These higher energies can be considered higher harmonics of the lower ones. How does intersystem crossing occur between two wave functions or states? How do we solve this problem? What does fine and hyperfine splitting of energy levels have to do with the observed sensitivity to NEMR?

The transition from one state to another, in quantum mechanics, is based on the spectroscopic selection rule, the transition moment integral of the two wave functions, and μ, the transition moment operator (the transition dipole moment), as expressed in Eqs. 5.11, 5.12, and 5.13.

Equation 5.11: Transition dipole Moment Change of State

$$|\Psi a > \text{to} |\Psi b >$$

$$\mathbf{(\Psi a|r|\Psi b) = (\mu / (E_b - E_a)_m)(\Psi n|p|\Psi b)}$$

$$\mathbf{E_a \text{ and } E_b \text{ equal the respective states' energy levels.}}$$

$$\mathbf{p = \text{momentum operator}}$$

Equation 5.12: Momentum Operator in One Dimension

$$\hat{p} = -i\hbar \frac{\partial}{\partial x}$$

Equation 5.13: Transition Dipole Moment Function in 3 Dimensions

$$\mathbf{(t.d.m)a \rightarrow b =< \Psi b|(qr)|}$$

$$\mathbf{\Psi a >= q \int\int\int \Psi b^{*}(r) r \Psi a(r)(dxdydz) r}$$

The integral (Eq. 5.13) is equal to the probability of the transition between the two states. If the integral is equal to zero, then the transition is "forbidden". The integral does not need to be calculated if the symmetry of the transition dipole moment function extends over all the point groups (states) to which the atom or molecule belongs. Then the integral's value is not zero, and the transition is allowed. A point group is a mathematical group (symmetry operation) that has a fixed point in common. The origin of a Cartesian system (within a Euclidean space) is a fixed point, and every point along the axes is in a subgroup of the point group.

The transition moment integral is zero if the transition moment function is antisymmetric or odd. The transition moment function symmetry or antisymmetry can be represented in Eqs. 5.14 and 5.15.

Therefore, if *R* is a relation imposed on a set A, then it is symmetric if relation Eq. 5.14 is met.

Equation 5.14: Relative Symmetry (Interchangeable Set)

$$\mathbf{R\ is\ a\ relation\ on\ set\ A.}$$

$$\mathbf{xRy \rightarrow yRx}$$

$$\mathbf{all\ x, y \in A\ then\ it\ is\ symmetric,}$$

$$\mathbf{x\ and\ y\ are\ interchangeable.}$$

For all *x* and *y* in set A, *y* can be exchanged for *x* in position.

R is antisymmetric if the conditions in relation to Eq. 5.15 are met.

Equation 5.15: Antisymmetric Versus Symmetric Sets

$$\mathbf{If\ xRy\ and\ yRx \Rightarrow x = y}$$

$$\mathbf{For\ all\ x, y \in A}$$

$$\mathbf{xRy\ then\ yRx\ is\ antisymmetric.}$$

Example:

$$\mathbf{R_1 = \{(1,1),(1,2),(2,1)\}\ is\ symmetric.}$$

$$\mathbf{R_2 = \{(2,1),(3,1),(3,2),(4,1),(4,2),(4,3)\}\ is\ antisymmetric.}$$

This relation indicates that $x = y$ for all x and y in set A; that is, if $x\ R\ y$, then y has no relation to x for all pairs, where a relation is suggested. Another way to state these set relation rules of symmetry/antisymmetry is any relation R on a set A is symmetric if $(a, b) \in R$ then $(b, a) \in R$ or antisymmetric if $(a, b) \in R$ and $(b, a) \in R$, then $a = b$. In other words, $(a, b) \in R \Rightarrow (b, a) \in R$ for symmetric and $(a, b) \in R$ and $(b, a) \in \mathrm{R} \Rightarrow a = b$ for antisymmetric. These mathematical exercises are useful because they provide a shortcut around a difficult calculation to determine the conditions under which a state change can or will occur.

In the case of DALM, will absorbed photons lead to immediate fluorescence or be stored, or will microwave or thermal energy facilitate the photonic energy storage or lead to its emission? Will free radicals be formed, or will the energy be converted back to photonic emissions through state changes? The preceding discussion of theory provides direction to addressing these observations experimentally rather than blindly hypothetically. The theory will be validated or falsified by the experimental data. Theory should lead experimentation, but not drive or dictate what experimental data is acceptable. If the data says otherwise, then the theory needs to be modified or changed altogether.

To this point, all this discussion on mathematical physical-chemical theory may seem academic and slightly overwhelming to molecular biologists and microbiologists, but a brief example will illustrate the benefits of understanding theory when optimizing the applications to be used by the latter. Early in the counterbiological warfare development of DALM (2005), experiments using continuous exposure to small to medium volumes of surrogate biological agent aqueous suspensions (*Bacillus thuringiensis* stand-in for anthrax spores) to demonstrate neutralization dramatically showed nonthermal effects. The conditions were reported in a subsequent engineering patent (thus the imperial system of measures). Initially, spore-killing experiments with 2450 MHz static exposures showed 10^4 killing in 2–3 min with and without DALM and its oxidatively activating reactants, but much of the microwave radiation was reflected, and heating to boiling occurred. This effect was evidently thermal. At first, when this system was replaced by a continuous flow, continuous microwave exposure closed system and the spore carrying volume increased from 35 to 600 mL, the luminescence (blue) glowed brightly until it was exhausted, and no spore killing was observed in 12-min exposure times. When 600 mL volumes and 600 watts power were applied, with cooling to prevent boiling and steam release, total spore killing was observed in 45-min exposures. More cooling fixtures were added to the system. Preliminary experiments with the new recirculating, cooled system showed complete killing of 10^8 spores at 45 min, with the temperature held at 80 °F (26 °C). When the temperature was increased to 100 °F (37.8 °C) the spore killing rate dropped to 10^5 [6]. The luminescence of the DALM was dependent on temperature, flow rate, and the presence of spores. Flow rates of 3 gallons (11.355 liters) per minute and temperatures at 75 °F (23 °C) did not

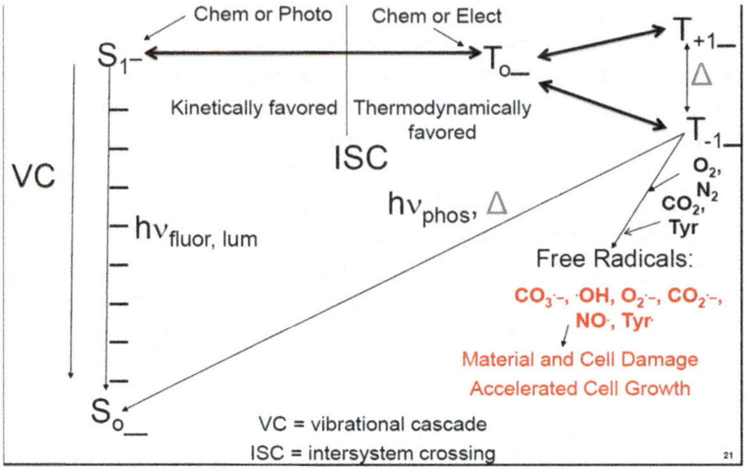

Fig. 5.4 The Jablonski diagram illustrating classical kinetic and thermodynamic interactions with the photochemistry and free-radical chemistry of DALM

produce a visible glow. Decreasing the flow rate to 0.05 gallon (0.18925 liter) per minute led to luminescence that was observed for over an hour without termination.

These data clearly demonstrated that conversion to radiative singlet states led to no killing, in contrast to nonradiative states, which subsequently converted to free radicals and killed spores. The quenching of luminescence by the spores indicated the trapping of triplet to doublet (free-radical) states, so they could not revert to singlet states.

The three electromagnetic field effect categories enumerated previously can best be illustrated in Fig. 5.4.

Increasing the lifetime of radicals resulting from singlet to triplet intersystem crossing accounts for a decrease in luminescence and an increase in microbial free-radical killing by DALM. However, the latter cannot fully explain the efficiency of the antimicrobial activity, especially when low-duty factor, high-pulsed power is applied.

The apparent "lossless" transition from singlet to triplet states appears to defy the second law of thermodynamics by not gaining or losing energy and heat, or does it (Figs. 5.5 and 5.6)? The "lossless" crossing of states through the "forbidden" barrier requires heating or microwave absorption simultaneously with specific photonic excitation. The quantity of heat or microwave energy absorbed is equivalent to that which drives the triplet states to singlet photon emitting states. The reverse transition from singlet to triplet nonradiative states releases the same amount of thermal energy. Therefore, this completely reversible transition appears "lossless." In the quantum theory just explored, the singlet and triplet states must be at the same energy level to make the transition. This can only be explained by quantum theory; classical electromagnetic theory fails. Cooling during the microwave exposure traps

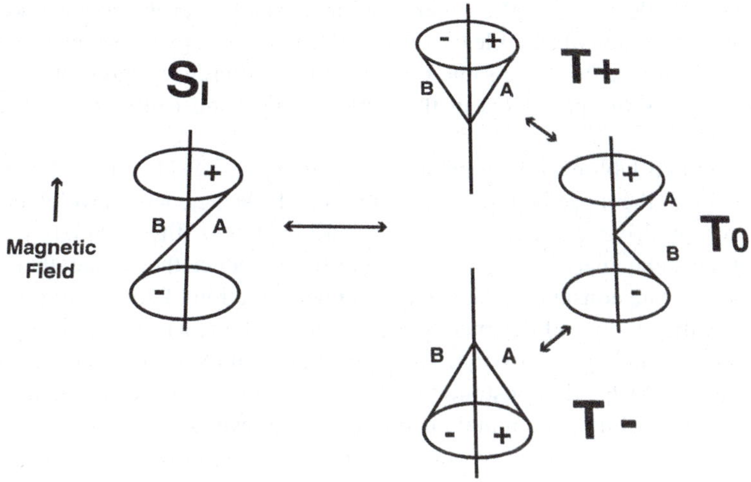

Fig. 5.5 A and B Precessional spin transitions of state

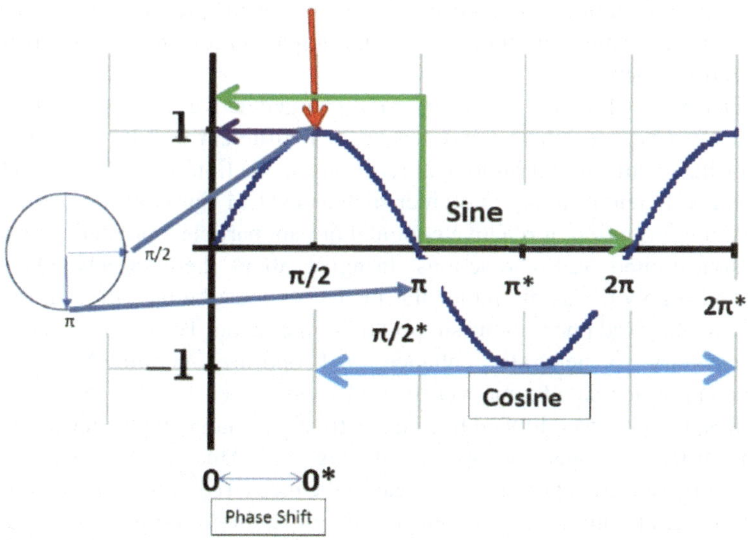

Fig. 5.6 The process is visualized as a magnetic trigonometric phase shift, a different initial condition, or a change in time

triplets that progress into doublets, free radicals, that cannot transition back into radiative singlet states. The triplets can only transition back into singlets if the same energy level is maintained at the same energy level as that of the singlets through microwave absorption (heating). In Chap. 7, I will address the role of entanglement in the "lossless" transition. Although these spin states are depicted as physical directions of spins about an axis, they are really the measure of magnetic susceptibility differences of electrons at the otherwise same energy levels. Since due to the Heisenberg uncertainty principle two electrons cannot occupy the same energy level in an atom or molecule at the same time, then this difference represents small discrete differences of spin states in the same energy band (differences in angular momentum).

The modified Jablonski diagram shows that two physical chemistry mechanisms can arise. First, there can be more triplet states at the same energy level than singlet states; entropy drives the excited states toward triplets. If these triplets are maintained at the same energy level as the singlet from whence they came, then even if the first-order rate constant is the same for either direction of conversion (perfectly reversible), the number of degenerate triplets drives the reaction kinetically toward the singlet state and light emission. The phenomenon of thermally activated delayed fluorescence (TADF), demonstrated by DALM, supports the requirement that the triplet energy level must be maintained at the same level as the corresponding singlet to undergo intersystem crossing. Furthermore, this observation suggests that the triplet states can decay and lose energy by nonradiative means, including decaying into stable free radicals. The application of magnetic fields favors the fine and hyperfine splitting of the energy levels of the triplets, increasing their energy separation and entropic number. This process also increases the production and lifetime of free radicals, allowing them more time to participate in recombination, abstraction, and reduction reactions.

The electric field polarizes and separates charged active species, including ionic free radicals. However, these processes are not sufficient to fully explain the efficiency of the chemical and biological reactions and actions, respectively. The poor diffusion coefficient of a large polymer such as DALM (into millions of molecular mass and the formation of reactive colloidal or nanoparticles) precludes rapid, even if diffusion-limited, redox reactions, bringing about these effects. Other more mobile and highly diffusible molecular mediators must be operating in these processes. The observed phenomena support this conclusion. The cofactors and coreactants necessary to generate, enhance, and optimize certain NEMR-induced phenomena provide data to uncover the underlying processes (Fig. 5.7).

The flashing observed looked like an electrical discharge in plasma and extended from the surface as a gas discharge might (Fig. 5.8). This is not a result of boiling and releasing ionized vapor and gas because the measured liquid temperatures were below boiling, although the gas temperature above the liquid meniscus was above that of the liquid (Fig. 3.6). An examination of the light emissions from these pulsing gases (bubbles in the liquid) and the discharges spectroscopically indicated that the gas was at temperatures comparable to the surface of the sun (10,000 K).

Additionally, the emissions (spectroscopy), as shown in the graph (Fig. 5.9), were of discrete wavelengths and very sharp peaks, not broadened by the presence of water vapor compared to laser-induced breakdown spectra [7]. Biological evidence supports the production and utilization of ionized and free-radical gas and plasma at high electronic temperatures. The generation and buildup of sound and its intensity with these flashes suggested a sonoluminescence mechanism.

Fig. 5.7 The most dramatic example, "lightning in a bottle," is produced by very brief, high-powered microwave pulses applied to dilute solutions of DALM containing luminol, low-concentration hydrogen peroxide, and a source of carbon dioxide: sodium carbonate, sodium bicarbonate, or carbonated water. As noted earlier, this flash was preceded by a building-up pulsing sound and light emissions until a flash discharge occurred

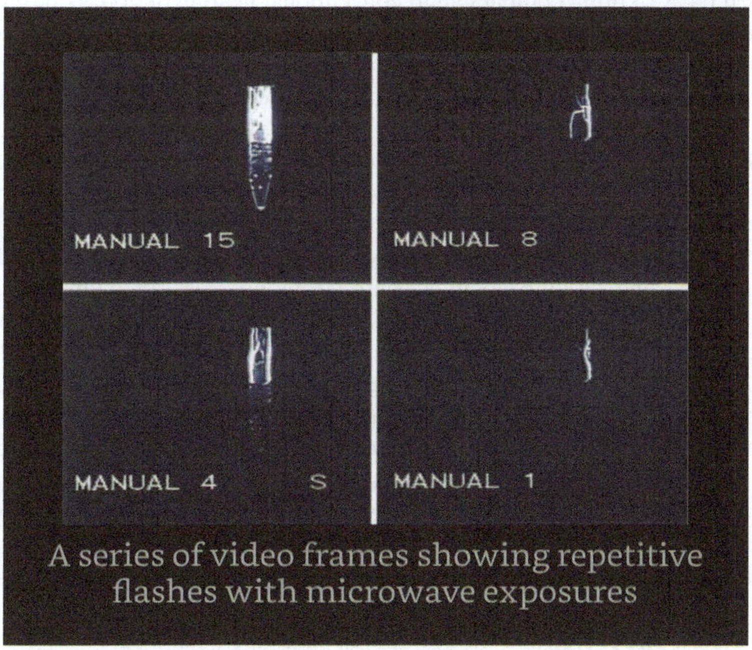

Fig. 5.8 A plasma discharge above the activated DALM mixture and luminous gas bubbles within the solution

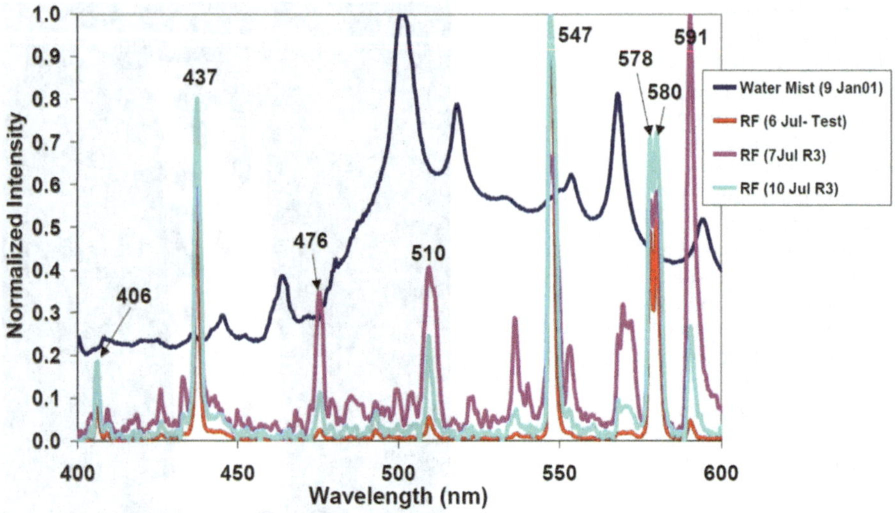

Fig. 5.9 Emission spectra of gas in liquid (water) of activated DALM solution compared to that of laser breakdown spectra. Note that the RFR pulsed spectra have narrower bands than that of the laser, lacking the spectral broadening of gas mixtures in water and looking more like those expected of gases at high temperatures (plasma)

Studies comparing the pulsed microwave killing of spores with DALM mediation to rapid conventional heating showed, even though microwave exposure never exceeded 64 °C, killing comparable to conventional heating of 200 °C or greater in 1 s (Figs. 5.10 and 5.11) [8]. These high temperatures mirror those produced by sonoluminescence [9, 10].

This leads to another general quantum mechanism that must be considered before we proceed—quantal absorption into coupled vibrational and rotational (degrees of freedom) states of gases [11].

Gases (diatomic) can be modeled as two masses separated by a spring (for vibrations) or a fixed rod (for rotations). The coupled version would be for a rotating unit separated by an oscillating spring or fixed rod governor with hinged rods with masses at the end circling about a central rotating axis. As with macroscopic mechanical governors, these would rotate fast until the masses began to rise or separate, respectively, and the law of conservation of angular momentum would slow their angular velocity, allowing them to approach each other again, again accelerating. This process of coupling would produce an oscillation of speeding up and slowing down. The vibrational states, separated by larger energy levels than rotational states, would be coupled to the rotational states. These rotational states are energetically close together (1–10 cm^{-1} difference), composed of rotational transition loops that couple to the vibrational transitions. The rotation is affected by its vibrational transition due to a change in bond length. Because vibrational energy states are in the range of 1000 cm^{-1}, rotational energy states are superimposed on vibrational energy states. The energies of the vibrational states are quantized.

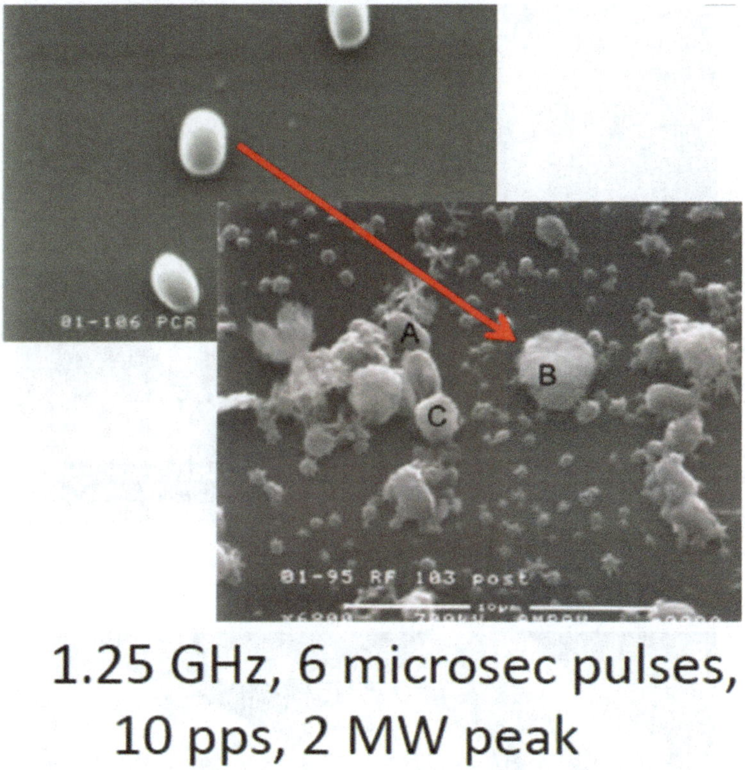

Fig. 5.10 Bacterial spores before and after pulsed microwave exposure (1.25 GHz, 6-microsecond pulse width, 10 pulses per second, 2 MW peak power) rapidly expanded and exploded

The symbol *u* is the vibrational quantum number with integer values of 0, 1, 2…n, and *v* is the frequency of the vibration. That frequency is given by Eq. 5.16.

The symbol *k* is the force constant, and μ is the reduced mass of the diatomic molecule with atom masses designated as *m*1 and *m*2 (Eq. 5.17).

The rotational quantization is given by Eq. 5.18.

Equation 5.16: Energy of Vibrational Quanta

$$\mathbf{E_v} = \mathbf{hv}\left(u + \tfrac{1}{2}\right)$$

Equation 5.17: Frequency of Vibrational Quanta

$$\mathbf{v} = \mathbf{1 / 2\pi\sqrt{k}} / \mu$$

$$\mu = \frac{m_1 m_2}{m_1 + m_2}$$

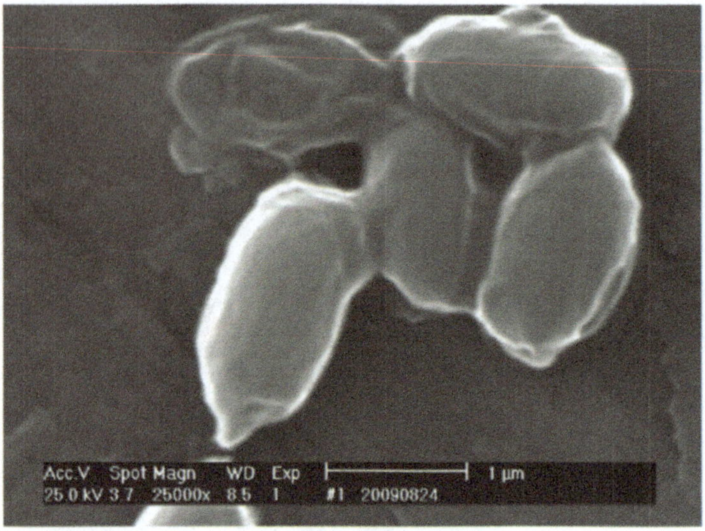

Control B.t. spores

B.t spores post non-thermal plasma treatment.

Fig. 5.11 Cold-plasma-exposed spores showed that they were "deflated" in a similar fashion by fenestration but less violently and immediately

Equation 5.18: Rotational Quantization

$$\mathbf{E_r = h^2 / \left(8\pi^2 I\right)]J\left(J+1\right)}$$

$$\mathbf{I = \mu r^2}$$

Equation 5.19: Function of Vibrational/Rotational Energy Level Change and R Branch

$$\mathbf{F\left(J\right) = E_r / h = \left[h / \left(8\pi^2 I\right)\right] J\left(J+1\right) = BJ\left(J+!\right)}$$

$$\mathbf{R - Branch}$$

$$\mathbf{\Delta E = hv_0 + 2hB\left(J+1\right)}$$

The moment of inertia is given by I, μ is the reduced mass, r is the equilibrium bond length, v_o is the vibrational frequency, and B = the rotational constant. Transitions in energy levels abide by selection rules. Two such selection rules lead to the transition $\Delta J = 0$ (i.e., $J'' = 0$ and $J' = 0$, but $v_o \neq 0$ is forbidden). Therefore, pure vibrational transitions usually do not occur. The rotational selection rule generates two branches of energy change: the R branch (when $\Delta J = +1$) (Eq. 5.19) and the P branch (when $\Delta J = -1$) (Eq. 5.20).

Equation 5.20: P-Branch Energy Level Transition of Rotational States

$$\mathbf{P - Branch}$$

$$\mathbf{\Delta E = hv_0 - 2hBJ}$$

Equation 5.21: Q-Branch Energy Level Transition of Rotational State

$$\mathbf{Q - Branch}$$

$$\mathbf{\Delta E = hv_0}$$

Equation 5.22: Total Spectra (Energy) of Vibrational and Rotational States

$$\mathbf{S\left(\mathit{u},J\right) = G\left(\mathit{u}\right) + F\left(J\right)}$$

Equation 5.23: Spectra of Combined Rotational and Vibrational States Based on Their Respective Quanta

$$\mathbf{S}(u,\mathbf{J}) = v_0\left(u + \tfrac{1}{2}\right) + \mathbf{BJ}(\mathbf{J}+1)$$

When $\Delta J = 0$, the ground state rotational quantum number is equal to the rotational quantum number in the excited state. This pathway is called the Q branch (Eq. 5.21).

It is in polyatomic molecules and diatomic molecules with electronic angular momentum in the ground electronic state, e.g., nitric oxide (NO). Most diatomic molecules, such as O_2, have a small moment of inertia, resulting in very small angular momenta and no Q branch. The total coupled energy, and therefore spectra, can be represented by Eqs. 5.22 and 5.23.

G(*u*) represents the energy of the harmonic oscillator, and *F*(*J*) represents the rotational energy. These gases are generated on the surface of the polymer and must be involved in the transfer of electron excitation energy from the polymer to the gases to the chemical or biological targets. This transfer of electronic energy, temperature, may result in sonochemiluminescence. The vibrational/rotational transfers in degrees of freedom are also involved in the photochemical and electrochemical behaviors of DALM, in which ionic radicals (negative charges and "holes," positive charges) may act as gases confined within the polymer chains.

In summary, the physical-chemical behavior of DALM in microwave fields is a totality of electronic, vibrational, and rotational quantum effects imbedded in classical electromagnetic dynamics. They address fields' impact on waveforms. Modern quantum field theory considers electrons and quarks packets of electromagnetic waves. Therefore, the quantum effects of electromagnetic fields are only on a different scale than the classic electromagnetics of RFR bioeffects [12].

Aside from gases and plasmas being involved in the complex behavior of the DALM composites, they demonstrate some unique nanoparticle behavior resembling that of quantum dots. This is the lead-in for the next chapter's engaging nanotechnology.

References

1. Program on Technology Innovation: Environmental and Health Issues Related to Radiofrequency Emissions from Smart Technologies: Summary of Two Workshops, EPRI Project Manager G. Mezei D. Dorr R. Kavet, Electric Power Research Institute (EPRI), 1024737, Final Report, December 2011.
2. Kiel, J. L., Type-b Cytochromes: Sensors and Switches, CRC Press, Boca Raton, Florida, 1995.
3. Zoltowski, B., et al., Structure of Full-length Drosophila Cryptochrome, Nature 480(7377): 396–399. doi:https://doi.org/10.1038/nature10618, 2011.
4. Yang, Z., et al., Cryptochromes Orchestrate Transcription Regulation of Diverse Blue Light Responses in Plants. Photochemistry and Photobiology 93: 112–127, 2017.

5. Liu, H., et al, The action mechanisms of plant cryptochromes. Trends Plant Sci. 2011 December; 16(12): 684–691. doi:https://doi.org/10.1016/j.tplants.2011.09.002.
6. Elson, J., and Lane, D., METHOD AND APPARATUS FOR DESTRUCTION OF BIOLOGICAL AND CHEMICAL AGENTS, US Parent 8.236,237B2, Aug.7,2012.
7. Kiel, J., Rockwell, B., Sutter, R., Williams, J., Hardin, D., Morales, P., Eikum, D., Thomas, R., Noojin, G., Alls, J., Seaman, R., and Mathur, S. Laser and Microwave Induced Breakdown Spectroscopy: Basis for a New Detection Technique for Chemical and Biological Agents, In Digest of Technical Papers, Pulsed Power Plasma Science 2001(R. Reinovsky and M. Newton, eds.), IEEE, New Jersey, pp. 220–223, 2001.
8. Kiel, J., Parker, J., Morales, P., Alls, J., Mason, P., Seaman, R., Mathur, S., and Holwitt, E. Pulsed microwave induced bioeffects. IEEE Transactions on Plasma Science 28(1): 161–167, 2000.
9. Glanz, J. The spell of sonoluminescence. Science 274: 718–719, 1996.
10. Barber, B.P., Hiller, R.A., Lofstedt, R., Putterman, S.J., and Wenninger, K.R. Defining the unknowns of sonoluminescence. Phys. Rep. 281: 65–143, 1997.
11. 13.2: Rotations Accompany Vibrational Transitions, LibreTexts, https://batch.libretexts.org/print/url=https://chem.libretexts.org/Bookshelves/Physical_and_Theoretical_Chemistry_Textbook_Maps/Physical_Chemistry_(LibreTexts)/13%3A_Molecular_Spectroscopy/13.02%3A_Rotations_Accompany_Vibrational_Transitions.pdf.
12. Strassler, Matt. Waves in an Impossible Sea: How Everyday Life Emerges from the Cosmic Ocean. Basic Books, 2024. 384 pp.

Chapter 6
Required Characteristics of a Nanostructure Radiofrequency and Microwave Absorbing Vector

Abstract The engagement of nanotechnology with classical electromagnetics and quantum mechanics imposed on nanometer constraints will be discussed here. The emphasis is on nanoscale effects within the previously discussed boundaries of classical and quantum mechanisms. This chapter serves as a prelude to Chap. 7, where all these components will be brought together to emphasize the sum of the isolated mechanisms and lead into the applied science and engineering of bionanotechnology radio-controlled vectors.

Keywords Nanotechnology · Nanoparticles · Carbon nanotubes · RNA · Electroporation · Viruses · Nanoradio

The importance of the phenomenon of diazoluminomelanin (DALM) interacting with metallic nanoparticles should not be understated. Unfortunately, it came at the end of the Air Force Research Laboratory's research program on bio counterproliferation. The graph in Fig. 6.1 illustrates the success of releasing the material from the control of introduced chemical reactants, except for water, and relinquishing it to physics and radio control.

What is the origin of nanotechnology's contribution to this overall process? We shall examine the redox chemistry and nonionizing electromagnetic radiation (NEMR) physics that led to the observable but heretofore unexplained solution.

The earliest and most practical application of electromagnetics to cell and molecular biology was electroporation (1980s). It grew out of previous decades of work on dielectrophoresis and electrorotation and subsequently included the magnetic manipulation of paramagnetic nanoparticles. Electrorotation arises from a phase lag between an applied rotating electric field and the corresponding relaxation processes. Dielectrophoresis is a force exerted on a dielectric particle when subjected to a nonuniform electric field [1]. This force ($\mathbf{F}_{\mathrm{Dep}}$ or time constrained $<\mathbf{F}_{\mathrm{Dep}}>$) does not require the particle to be charged (Eqs. 6.1 and 6.2).

J. L. Kiel, *Bionanotechnology*, https://doi.org/10.1007/978-3-031-63006-4_6

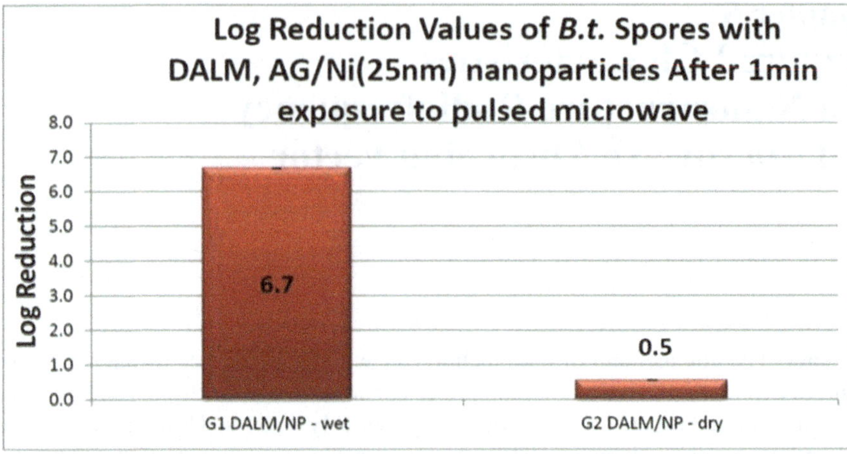

Conditions: 9.02 GHz, 50 kW peak power, 1μs pulses, 500Hz pulse frequency, 1 minute exposure; Note that no activating solutions of hydrogen peroxide, luminol, and/or sodium bicarbonate were required for this high efficacy kill, only water (very limited amount) and the composite nanobe, Nanobe OAJ7. No special exposure orientation was required; polarization effects have been overcome.

See U.S. Patents 7,892,484 (10 Jan 2012) and 8,383,036 (26 Feb 2013)

Fig. 6.1 The success of using DALM nanoparticles to kill bacterial spores activated only by water and pulsed microwaves

Equation 6.1: Dielectrophoresis Force for a Sphere

$$\mathbf{F}_{\mathbf{Dep}} = 2\pi \mathbf{r}^3 \varepsilon_{\mathbf{m}} \mathbf{R}_{\mathbf{e}} \left\{ \left(\varepsilon^*_{\mathrm{p}} - \varepsilon^*_{\mathbf{m}} \right) / \left(\varepsilon^*_{\mathrm{p}} + 2\varepsilon^*_{\mathbf{m}} \right) \right\} \nabla \left| \mathbf{E}^{\rightarrow}{}_{\mathbf{rms}} \right|^2$$

Equation 6.2: Dielectrophoresis Force for an Ellipsoid

$$\mathbf{F}_{\mathbf{Dep}} = \left(\pi \mathbf{r}^2 L \right) / \mathbf{3}) \varepsilon_{\mathbf{m}} \mathbf{R}_{\mathbf{e}} \left(\varepsilon^*_{\mathrm{p}} - \varepsilon^*_{\mathbf{m}} \right) / \varepsilon^*_{\mathbf{m}}]\} \nabla \left| \mathbf{E}^{\rightarrow} \right|^2$$

$$\nabla = \hat{x}\frac{\partial}{\partial x} + \hat{y}\frac{\partial}{\partial y} + \hat{z}\frac{\partial}{\partial z}$$

where "R_e" indicates that the real part of the expression is used: E·ejωt = E(cos ωt + j sin ωt), so that R_e[E] = E cos ωt. ε^*_p = particle permittivity and ε^*_m = medium permittivity.

Electroporation is the application of high-voltage pulses of an electric field to cells to increase the permeability of their membranes to allow chemicals, drugs, electrode arrays, or deoxyribonucleic acid (DNA) or messenger ribonucleic acid (mRNA) to enter the cells and reseal them [1]. These processes have become an important part of developing a radio-controlled vector. As mentioned earlier, high-aspect-ratio carbon nanotubes have been shown to interact with microwave fields by concentrating E fields at their tips (lightning rod effect), which facilitated their penetration of bacterial cells. A magnetron device exposure at 100 W, 2.45 GHz, for 1 s was used and repeated at different MW exposures of 2–10 s [2]. The investigators showed attachment of their carbon nanotubes and gold nanoparticles to the bacterial cell membranes by SEM (scanning electron microscopy) but did not show internalization in these preliminary studies (no transmission electron microscopy (TEM)) [3–7].

The extrapolation from the attachment of nanostructures to cell membranes to observable electroporation, by the authors of this study, for the method to replace the long-established one for electroporation, DC high voltage pulses, to my knowledge, as a wireless alternative, has never come into fruition, although it is still being rigorously investigated. This book is, in part, about how this goal was finally accomplished.

The nature of the shape of nanostructures with respect to radiofrequency radiation (RFR) interaction was even raised due to the COVID-19 pandemic. I participated in a federal government advisory group on whether RFR and microwave radiation could destroy SARS-CoV-2 in the air and whether pursuing research in this area would be worthwhile.

All this attention was based on a published, in part, theoretical paper with some preliminary data [8]. The modeling in this paper suggests that all the energy is absorbed in the viral particles (but water or even surrounding materials were not considered). The theory is based on the forces applied by the electric field being converted efficiently into vibrational energy to fracture the protein capsid and lipid envelope structure. This suggests vibrational resonance. Dampening and relaxation effects are missing, and the contribution of water is neglected. An RFR absorber requires semiconductor properties and structures to concentrate and amplify local electric fields to facilitate the conversion of changing electric fields into vibrations and sound (as previously stated, single-walled carbon nanotubes (SWCNTs)) can do this with high-aspect-ratio ends). If vibrations were the cause of NEMR interaction with viruses, they should “sing” or “chirp” with pulsed RFR if their concentrations are high enough and this mechanism is operational.

There is a nanostructure that does convert RFR into vibrations. It uses carbon nanotubes with dimensions of 10 nm by 100 nm. The nanotube “wire” must be suspended taunt in a vacuum of 10^{-7} mm mercury to vibrate with RFR. A multilayered carbon nanotube (MWCNT) was placed on a silicon electrode in a vacuum and was connected to a counterelectrode through a wire supplied by a direct current (DC) source. The CNT acts as an antenna, a tuner, an amplifier, and a demodulator, all in one. The CNT acts as an antenna by vibrating at the same frequency as the

RFR received. It acts as a tuner by the RFR frequency being optimally close to the natural frequency for the length of the CNT. It acts as an amplifier because of its high aspect ratio (again, the lightning rod effect). It acts as a demodulator by vibrating in sync with the carrier wave but responding only to the information in the signal component, not the carrier wave itself. The nanoradio was discovered by Sumio Iijima in 1991 [9].

In addition to the size and shape of nanostructures, is there anything about them in general that determines their chemical reactivity? The answer is found in the surface area of the nanoparticles. The same mass divided into many nanoparticles greatly increases the surface area. Most chemical kinetics, especially those important to biology, such as in aqueous solutions, are based on the first assumption that they occur in dilute solutions. In my graduate research, I realized that the enzymatic reaction I was investigating, and the redox killing of cells by activated peroxidase, could not be modeled by conventional enzyme kinetics with soluble small-molecule substrates. The cell surface was the substrate for the much smaller, more mobile molecules of peroxidase. To construct the appropriate kinetics model that was supported by actual data, I needed a totally different approach. The kinetic model, although largely empirical, was based on Einstein's model of molecules diffusing and sweeping out a given volume driven by Brownian motion (discussed earlier). I stepped back from the volume and considered the peroxidase cross section and the corresponding target areas on the surface of the cell, with the addition of lateral single molecular layer diffusion, which are the determining factors of the chemical reaction rate.

This model is a kind of "target theory." The limiting reaction step is striking the target, diffusion being limited, not postbinding reaction rates. Now applying this reasoning to nanoparticles, redox reactions involving the surfaces of such particles interacting with oxygen and, in turn, with cellular targets (at a much slower rate unless preplaced on the cell surface) would be congruent with this mechanism. An example of how this works is illustrated in Eq. 6.3 on peroxidase lysing red blood cells (RBCs). RBCs are more vulnerable to lysis than normal red cells if cholesterol is first extracted from the RBCs (single data point X in Fig. 6.2).

The overall reaction rate (rate-limiting step) is then related to the surface area (spherical surface area = $\mathbf{4\pi r^2}$) of the nanoparticles (Eq. 6.3).

Equation 6.3: Rate Equation for Red Blood Cell Lysis by Activated Peroxidase

$$r_L = kp\sqrt{[A]}\left[\frac{n4\pi r^2(P_f)}{N(PA)(V - V_c + V_m)(10^{-3})}\right] - \frac{2kd}{[E]kf}$$

(1) r_L = rate of lysis, cells lysed per second; (2) kp = $3.02 \times 10^{10}\ M^{-1}\ s^{-1}$ = diffusion-limited rate constant for the propagation reaction; (3) kd = rate constant for toxic

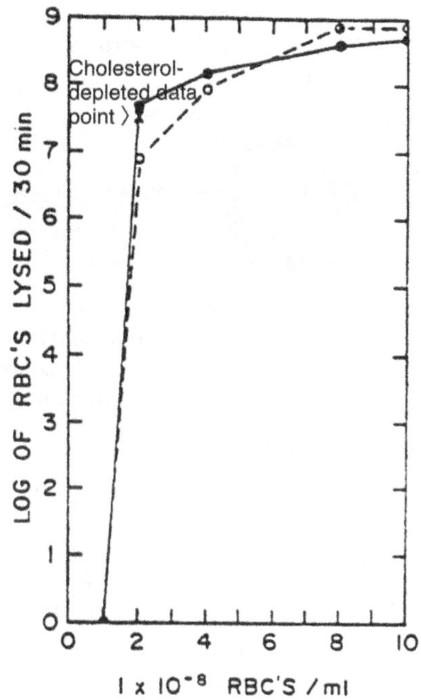

Fig. 6.2 The solid line is the calculated lysis results (with the exception of the single datum X of the RBCs with cholesterol extracted, marked on the graph with a pointer). The dashed line is from experimental hemolysis experiments

peroxidase decay, 1.39×10^2 s; (4) kf = rate constant for toxic peroxidase formation, $0.91 \times 10^2\ M^{-1}\ s^{-1}$; (5) A = mean concentration of the monomer in the cell membrane, 0.054 M; (6) E = total concentration of peroxidase, 1.25 μM; (7) n = number of cells in a 1-mL reaction volume; (8) r = radius of the cell in cm, 3.5×10^{-4} cm for the erythrocyte; (9) P_f = fraction of surface area of the cell accounted for by monomer (phospholipid) containing unsaturated fatty acid acyl groups, 0.068; (10) N = Avogadro's number, 6.02×10^{23}; (11) PA = surface area of the monomer (phospholipid), $0.52 \times 10^{-14}\ cm^2$; (12) V = suspension volume, 1 cm^3; (13) V_c = total cell volume in cm^3; (14) V_m = total membrane volume, n X ($6.16 \times 10\ cm^{-13}$). As V-V_c approaches 0, the entire rate law becomes rL = k [A] $^{3/2}$.

The combination of the size, shape, and composition of the DALM metallic nanoparticles was first assumed to explain the interactions with RFR and the microwaves observed (Fig. 6.3). The original triclinic crystal-like structure, which led to macroscopic ribbons with angular pointed ends, was thought to concentrate and amplify the pulsed RFR and convert it to sound, like the MWCNT in nanoradios. Instead of a battery, the electrons for amplification were provided by the ongoing redox reactions in the DALM polymer during RFR exposure. This straightforward classical explanation became further frustrated by the chemical and nanotechnology configurations that worked best.

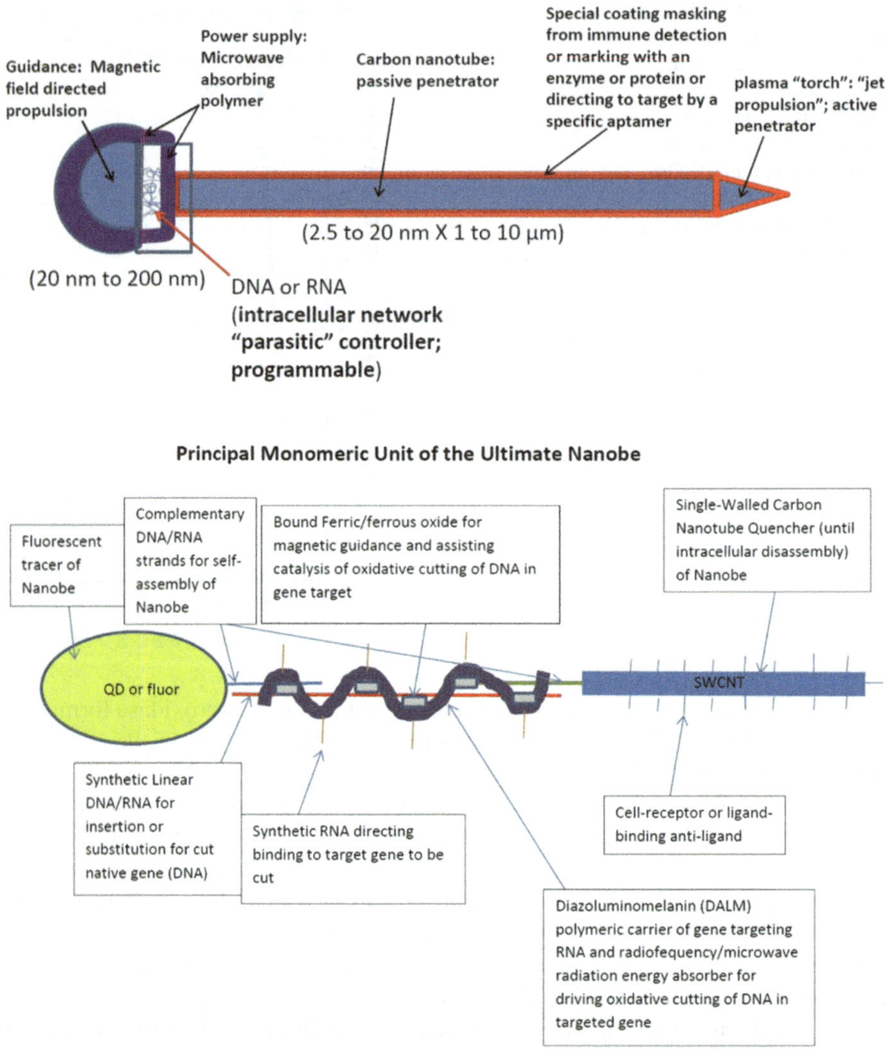

Fig. 6.3 Composite DALM nanoparticle designs for diagnostics, tracers, and antimicrobial and genetic modification functions

References

1. Jaffe, A. and Voldman, J., Multi-frequency dielectrophoretic characterization of single cells, Microsystems & Nanoengineering 4:23, 2018.
2. Giersig, M., et al., Novel Electroporation System for both Gram-negative and Gram-positive Bacteria Assisted by Multi-Walled Carbon Nanotubes, Mater. Res. Soc. Symp. Proc. 845: 285–290, Materials Research Society, AA5.21, 2005.

3. Kiel, J., Parker, J., and Bruno, J., (AF Inv 20,635) Enhanced Nitrite Production in Transfected Murine Cells. U.S. Patent 5,464,768 (issued 7 Nov 95).
4. Conde, J., et al., Gold-nanobeacons for simultaneous gene specific silencing and intracellular tracking of the silencing events, Biomaterials 34: 2516–2523, 2013.
5. Rojas-Chapana, J., et al., Enhanced Introduction of Gold Nanoparticles into Vital *Acidothiobacillus ferrooxidans* by Carbon Nanotube-based Microwave Electroporation, Nano Letters 4 (5): 985–988,DOI: https://doi.org/10.1021/nl049699n, 2004.
6. Wang, T., Brown, D., and Xie, X., Operando Investigation of Locally Enhanced Electric Field Treatment (LEEFT) Harnessing Lightning-Rod Effect for Rapid Bacteria Inactivation. Nano Letters 22 (2): 860–867. https://doi.org/10.1021/acs.nanolett.1c02240, 2022.
7. Gule, N., et al., Immobilized Furanone Derivatives as Inhibitors for Adhesion of Bacteria on Modified Poly(styrene-co-anhydride). Biomacromolecules 13 (10): 3138–3150. https://doi.org/10.1021/bm300932u, 2012.
8. Yang, S., et al. Efficient Structure Resonance Energy Transfer from Microwaves to Confined Acoustic Vibrations in Viruses. Sci Rep 5, 18030, https://doi.org/10.1038/srep18030, 2016.
9. Service, R., TR10: NanoRadio, MIT Technology Review, http://www2.technologyreview.com/news/409595/tr10-nanoradio/, February 19, 2008.

Chapter 7
Seeking Explanations for Paradoxical Responses: Bringing It All Together

Abstract This chapter is a turning point from the broad theoretical basis for fundamental chemical and biological macromolecule interactions with nonionizing electromagnetic radiation to bioengineering radio-controlled genetic and antimicrobial vectors, which hopefully will become powerful new tools for molecular biologists. The first step is reconciliation, at this point, of all theories discussed heretofore with the data observed. This chapter is a preparation for the next pragmatic and practical steps toward a broad range of applications.

Keywords Melanin · Globular · Bound water · Modified Jablonski diagram · Precession · Time · Phase

Molecular biologists and microbiologists may feel they have been patient to suffer through the physics, electrical engineering, and chemistry of the previous chapters, and physical scientists and engineers may say the same about what is to come, concerning molecular biology, but the success of the collaborations of the respective fields will make the necessity evident.

The first discord to resolve is the necessary interactions between classical electrodynamics and thermodynamics and quantum mechanics to yield the observed phenomena. We will start by revisiting the data already noted and adding more to visualize the unique properties of diazoluminomelanin (DALM) and those that might be shared with natural biological macromolecules as well and in combination with their smaller molecular substrates. Early data, using luminol, DALM, and DALM-coated magnetite ("fluidized magnetite": preparation with particles ranging from nanometer to micron diameters), showed remarkable differences in their responses to 2450 MHz, 25-watt, continuous wave microwaves (Fig. 7.1).

Luminol luminescence, with hydrogen peroxide alone, appeared to be very thermally sensitive above 31 °C, but in fact, this was a threshold for the "no slope" of the luminescence in response to temperature. Luminol luminescence became insensitive to temperature and continued to increase monotonically and linearly because the substrates were in excess (the rate-limiting step was no longer governed by temperature change). As noted in an earlier paper on diazoluminol, the

J. L. Kiel, *Bionanotechnology*, https://doi.org/10.1007/978-3-031-63006-4_7

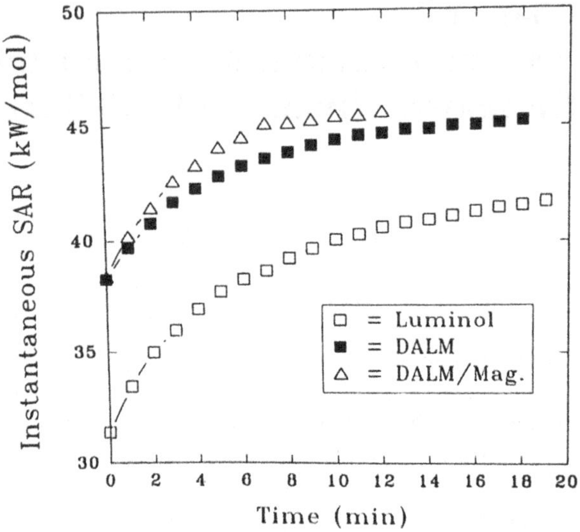

Fig. 7.1 The instantaneous SAR was dependent on the chemistry that changed the microwave absorption

thermochemiluminescence of diazoluminol was most stable starting at 0 °C (with carbon dioxide or carbonate) but became less stable at higher temperatures. In the same paper, carbon dioxide had to be flushed from the solution for the response to ultrasound to be stable. On the other hand, DALM showed a greater thermochemiluminescence response from 27 to 45 °C than luminol in an increasing curvilinear pattern. Carbon dioxide or carbonate enhanced this response, and additional luminol, with hydrogen peroxide, enhanced it but with the properties of the DALM and not free luminol. When observed in the time domain, the magnetite-bound DALM peaked luminescence rate faster than DALM or luminol alone. Both DALM and DALM magnetite produced far more luminescence than luminol. Oddly, when the thermal microwave responses of DALM and DALM magnetite were compared, the DALM magnetite was less thermally sensitive than DALM above 31 °C. However, to reconcile these two sets of responses, the DALM magnetite must have heated faster than the DALM.

The metal may provide one of two possible mechanisms. The first mechanism is the heating of a metallic object with surface E field reflection and the possible currents produced in the metallic material. The second mechanism is the ferromagnetic particles switching the states of the DALM from singlet, light-emitting to triplet, nonradiative, heat-producing states. Additionally, the finely divided formulation of the magnetite particles may have led to a pyrophoric metal that rapidly autooxidized and further oxidized the DALM, making it more susceptible to microwave effects. It is possible that both of these mechanisms contributed to the effect. The latter has been evident in the recent observations of micro/nanoparticle stability. When coated with DALM, they remain ferromagnetic and can be pulled from the solution. The uncoated particles "rust" and lose their ferromagnetic property and remain suspended in the solution.

Next, we consider the direct current (DC) electric field effects on DALM at a fixed temperature. Here, the importance of the isolated E field is emphasized. Two observations give a strong indication of the mechanism. First, the luminescence response is immediate to a DC square voltage pulse, correlating positively to the voltage intensity of the pulse, and appears to decay with the duration of the pulse train. Second, the signal appears to recover if there is sufficient time between pulses, and if not, it appears to deplete with each subsequent pulse. However, this is not because of the depletion of luminescent reactants but rather because the states are shifted into slow luminescing states that raise the baseline luminescence and sustain it between electrical square waves. Additionally, this interperiod steady-state baseline increased with subsequent pulses. This same building of luminescence, along with the immediate corresponding luminescence pulses, occurred with high-power, low-duty factor radiofrequency radiation (RFR) pulses. In addition, the latter were accompanied by corresponding sound pulses.

This sound production suggested other mechanisms, such as sonoluminescence [1–3] or electrical discharge heating. The sound production correlates with the pulse and not the carrier wave. It resembles the nanoradio responses (multiwalled carbon nanotubes (MWCNTs)) but without the vacuum and applied DC voltage (electrodes). However, the actual information conversion of RFR signals by the DALM nanoparticles into sound has yet to be accomplished. These two mechanisms are opposed in the pressures of gases required to express them. The first, sonoluminescence, requires generating vibrations that compress gas bubbles generated in the process by shock waves, creating pressures in the range of 300 bars. The second, electrical breakdown heating, requires the voltage breakdown of a material or, more likely, a low-pressure gas ionized by microwaves (plasma generation) that discharges the peak-induced voltage by the RFR or microwaves. Neither of these mechanisms satisfactorily and completely explain the data. The measurements of the temperatures of pulsed flashes spectroscopically indicated that these were very-high-temperature (10,000 °C plus) gas emissions, even though the bulk temperatures were well below boiling. As noted earlier, the killing of heat-resistant spores by continuous-wave microwaves with DALM nanoparticles actually required cooling that decreased luminescence, counterintuitively.

The high-power, low-duty factor pulsed microwaves killed heat-resistant spores with very short time exposures that caused killing rates comparable to 200C plus for 1 s. The lightning-like discharges above the reaction liquid, periodically during the exposure, with pulsed light between regular sound and light pulses synchronous with the microwave pulses, were suggestive of electrical discharge in plasma.

The radiative-nonradiative-state mechanism of DALM nanoparticles is not just about forming reactive small molecules or gases (plasma) to generate the observed effects but also the containment of active species within the conjugated structure of the DALM polymer. This latter part of the composite, the complex mechanism of DALM nanoparticle luminescence, redox catalysis, and cytotoxic effects, is supported by a phenomenon typical of inorganic nanoparticles, quantum dots (Fig. 7.2).

DALM demonstrates a quantum dot phenomenon called "blinking" after the power is turned off (Fig. 7.3) [4, 5].

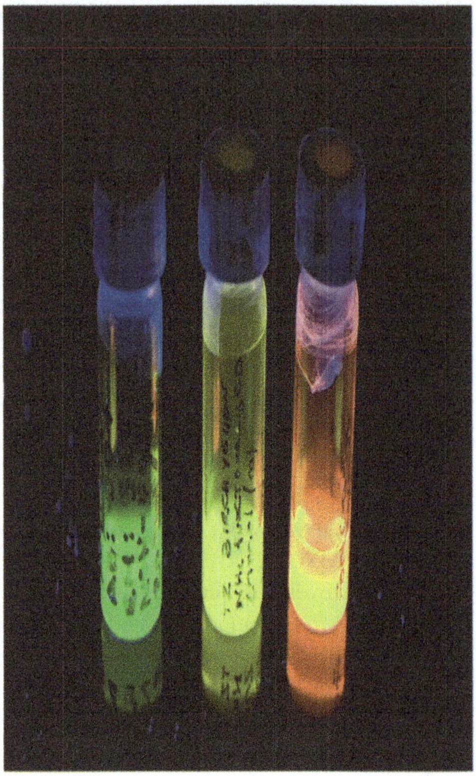

Fig. 7.2 Glowing quantum dots; the color of fluorescence depending on the size and not the chemical composition of the quantum dots

This is basically intermittent fluorescence (photoluminescence) or electroluminescence (as in light-emitting diodes (LEDs)) due to switching from a radiative to a nonradiative state. It is based on one of several mechanisms. The most often cited is the Auger effect, which is a type of electron and hole (in a semiconductor conduction band or valence band) recombination that does not produce light, but rather, the excess energy released is transferred to a second electron instead of generating a photon [6]. The excited electron transfers this excess energy to another electron that dissipates the energy by collisions with the crystalline lattice of the material, yielding vibrational energy decaying to the ground state of the band edge. It is a multiparticle interaction within a crystal lattice or polymer. The virtual particles can be multiple holes or actual electrons in the conduction or valence bands of the semiconductor. The DALM polymer conjugated carbon bond structure acts as the conduction band of a semiconductor.

Positive free radicals act as holes, and negative free radicals act as electrons, negatively charged carriers. The probability of Auger recombination depends on the density of these charge carriers and, thus, the rise in vibrational intensity (heating and/or sound generation) as their numbers increase with electromagnetic pulses and the rate of accompanying redox reactions [6]. There are two competing models to explain this blinking: Auger recombination, as discussed, and surface-trap-induced

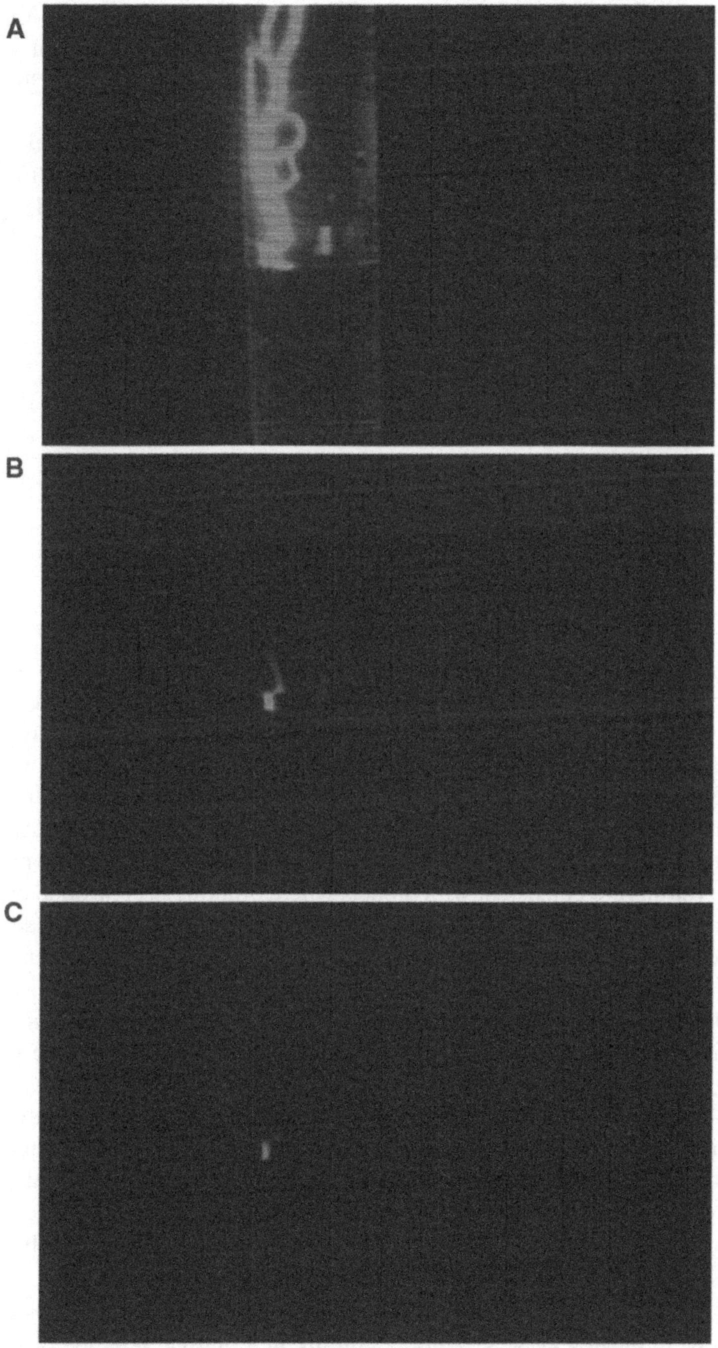

Fig. 7.3 (**a**, **b**, and **c**) A series of video frames of a "blinking" bubble trapped in a capillary tube, following the turning off of microwave pulses

recombination [6]. Auger blinking can yield exponential on/off times. A change in the power law statistics of the probability of fluctuations in emission intensities (distribution of light and dark states) indicates changes governed by millisecond time constants that constrain the quantum dot blinking mechanisms.

The observations involving quantum dots are consistent with models based on diffusion-limited energy transfers of electron or hole-trapped states. Blinking periods are described by probability densities, as in the equations in Eq. 7.1.

Equation 7.1: Probability Density of Quantum Dot Blinking

$$\mathbf{R_E = Bn^2p}$$

$$\mathbf{R_H = Bp^2n}$$

$$\mathbf{B = Auger\ coefficient}$$

$$\mathbf{T_E = p / R_E = 1 / Bn^2}$$

$$\mathbf{T_H = n / R_H = 1 / Bp^2}$$

The lifetimes (T = tau) of the blinking are inversely proportional to the squares of the negative (electrons) and positive (holes) in the material. The power law holds regardless of the temperature, quantum dot size or composition, nanoparticle shape, or excitation intensity. R_E and R_H are the recombination rates for electrons and holes, respectively. The power law is defined as a functional relationship between two quantities, where a change in one quantity results in a proportional relative change in the other quantity, independent of the initial size of those quantities.

Observations are consistent with models where blinking is controlled by the slow diffusion of the energies of electron or hole-trapped states. In such models, photoluminescence is quenched by the trapping of carriers by the trap state energetically meeting a resonance condition. The durations of blinking periods are determined by the diffusion of the energies of trap systems near the resonance. Diffusion-controlled mechanisms predict the existence of a critical time below which the power-law blinking statistics change.

With respect to DALM, the blinking indicates that DALM not only acts as a redox catalyst for small luminescent molecules but also participates in photochemical reactions, including thermochemiluminescence and polymer free radical formation. A higher-order structure of the polymer makes this possible.

The mechanisms I have discussed thus far can be summarized in the modified Jablonski diagram in Fig. 7.4.

Long-lived radicals (Fig. 7.5) must be more energized to become effective killing radicals, as in plasma gases causing "lightning in a bottle." When cold plasma gas

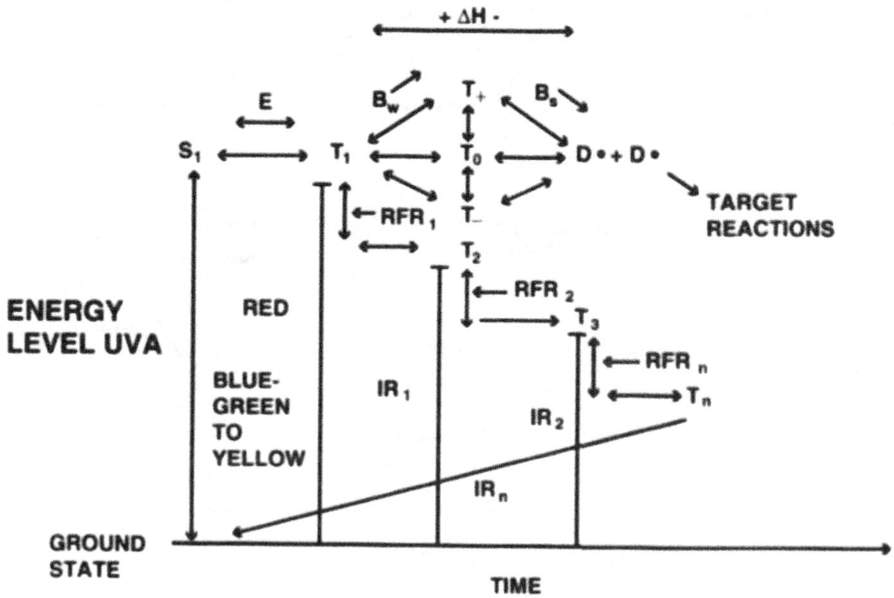

Fig. 7.4 Jablonski diagram showing the influence of magnetic fields and RFR on the distribution of activated states of DALM

Fig. 7.5 ESR measurement of the dry quiescent polymer indicates that it maintains stable, long-lived free radicals. The ESR spectrum is of dry DALM at room temperature (25 °C). A Varian E-line spectrometer was used at 9.48 GHz, 10 mW power, a field set of 3353 G, a scan range of 400 G, a modulation amplitude of 16 G, a time constant of 3 s, a receiver gain of 2×10^3, a modulation frequency of 100 kHz, and a scan time of 2 min

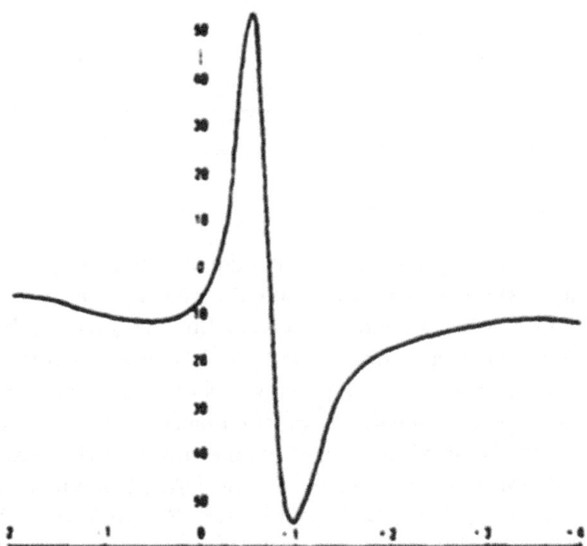

was applied to resting DALM, it quenched the plasma's killing power. Dry DALM showed more than baseline (but still weak) killing. The water inhibited rather than enhanced the killing power.

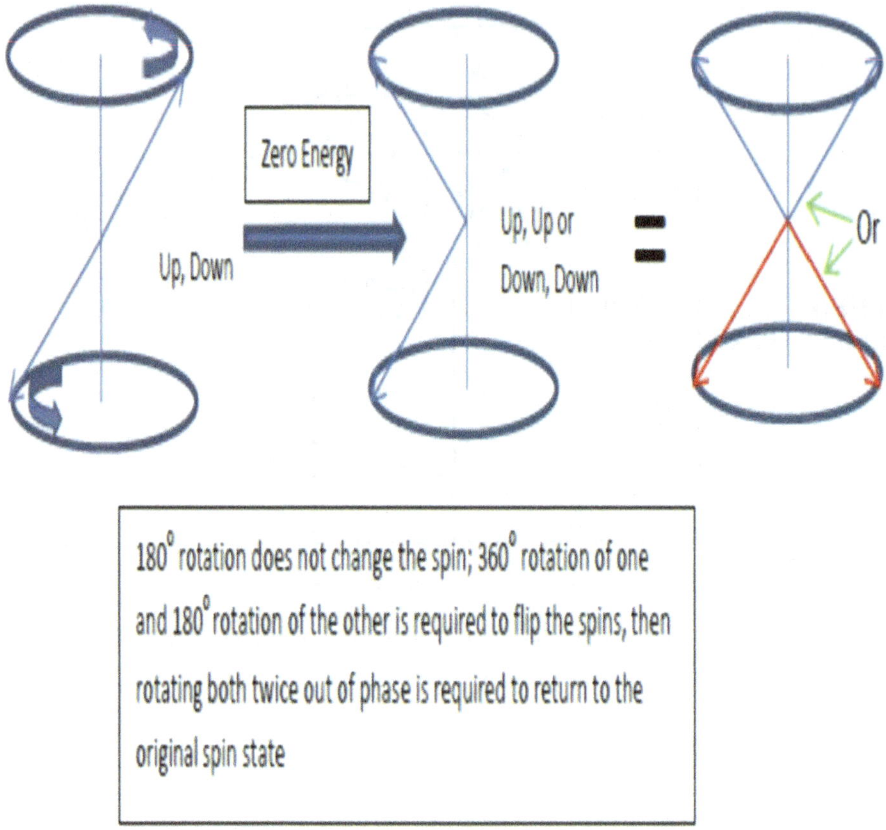

Fig. 7.6 Precessional changes that represent entanglement and lossless transitions of activated states

The precessional phase diagram shown earlier, along with the previous modified Jablonski diagram, shows how the transition from singlet to triplet states and back can be "lossless" and how a weak Lorentz force can bring it about.

The phase diagram shows how the "superposition" of the precessions represents entanglement, where the two states of equal energy are indiscernible because both times (phases) are displayed simultaneously (Fig. 7.6).

The linear, fibrous, and ribbon forms of DALM are supported by the high-aspect ratio microscopic structure of the DALM polymer, whether synthetic or initially biosynthetic (as synthesized by *Bacillus anthracis*). This linear structure with the activated state containing cyclic subunits rotates in and out of alignment at frequencies determined by energy absorption rates that overcome the barriers to rotation and in-plane alignment. These properties favor the maintenance of entangled but physically separated states.

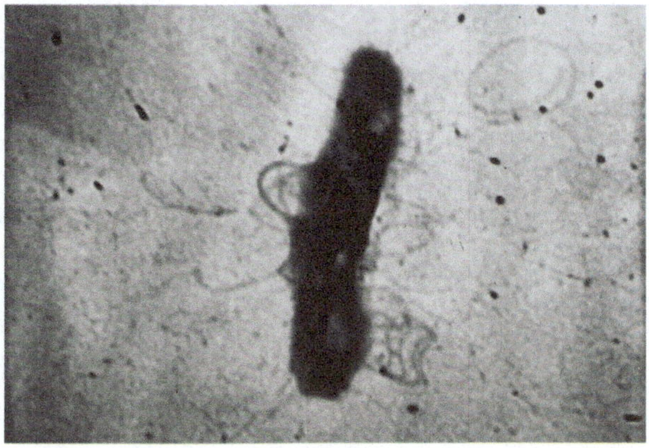

Fig. 7.7 Biosynthesized DALM extruding from a vegetative bacillus of *Bacillus anthracis*

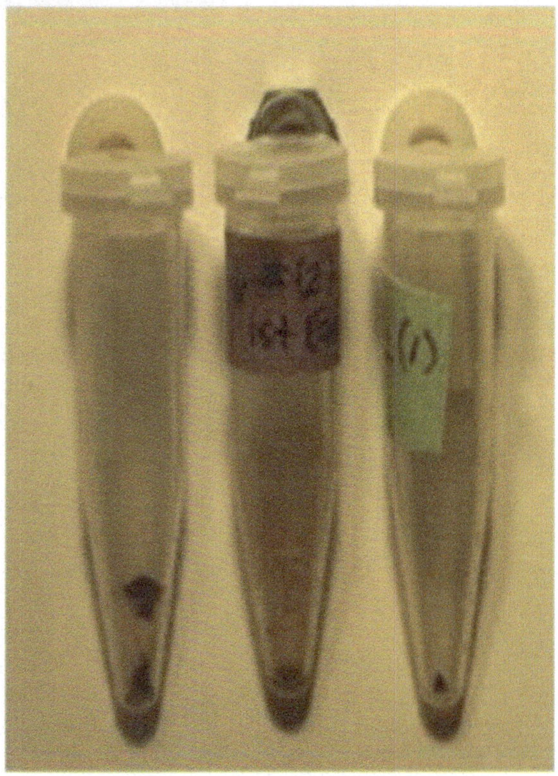

Fig. 7.8 DALM copurifying with the bacterial DNA upon extraction

The growth morphology of *Bacillus anthracis* on medium-inducing DALM biosynthesis was also interpreted as supporting fibrous DALM formation [7] (Figs. 7.7, 7.8, and 7.9).

In addition to the extruding fibers in the electron micrographs, the coiling of the *Bacillus* was evident when grown on a 3AT medium, especially when dilute

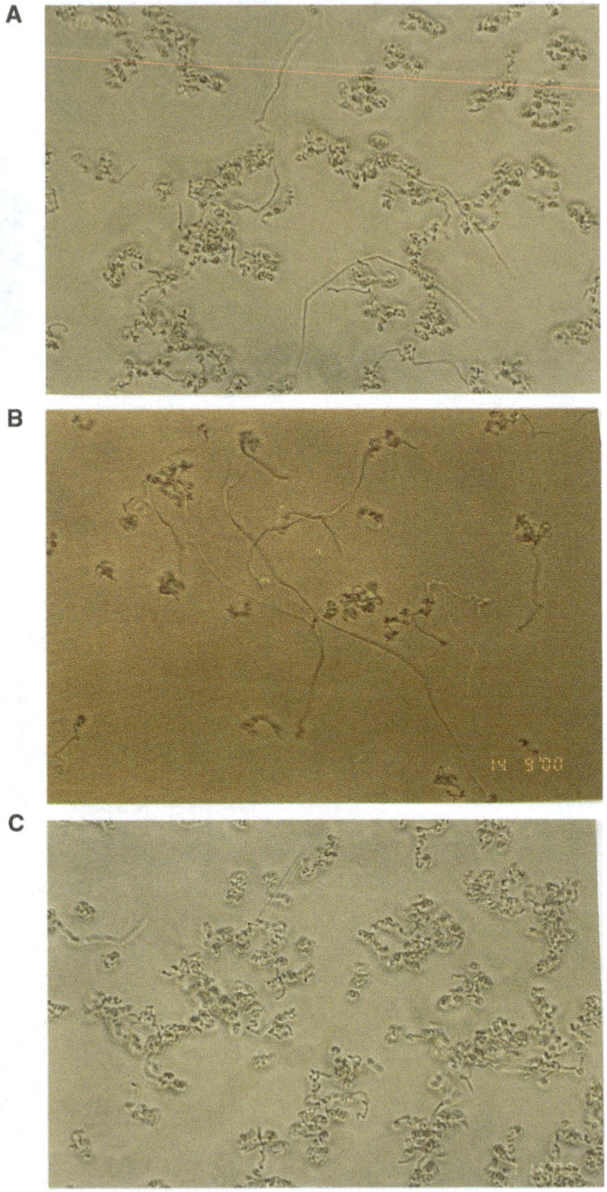

Fig. 7.9 (**a**, **b**, and **c**) Examples of the coiling and branching of *Bacillus anthracis*, vegetative anthrax bacteria, grown on a 3AT medium and enhanced by hydrogen peroxide and sodium carbonate

hydrogen peroxide and sodium carbonate were added. All variants of anthrax bacteria showed accelerated growth and coiling (under the right conditions) on a 3AT medium (Figs. 7.9, 7.10, and 7.11).

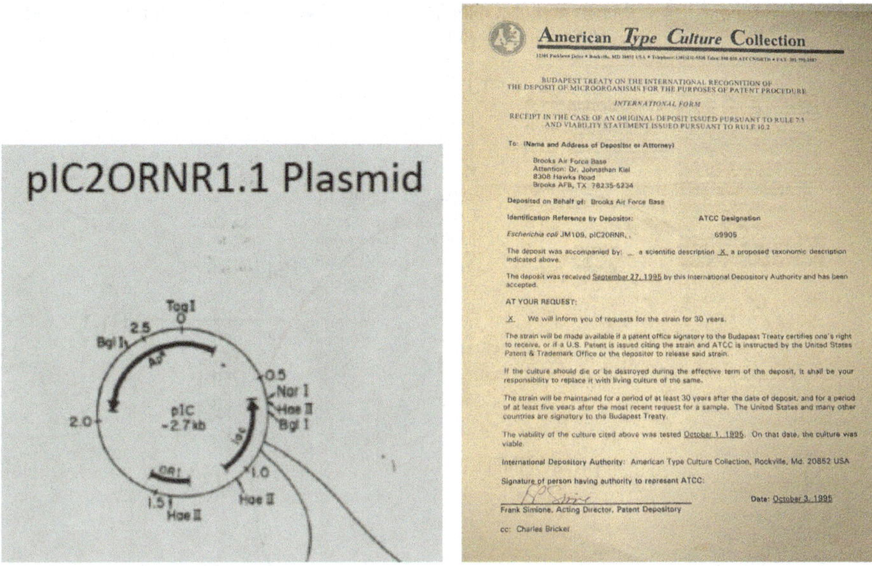

American Type Culture Collection

BUDAPEST TREATY ON THE INTERNATIONAL RECOGNITION OF THE DEPOSIT OF MICROORGANISMS FOR THE PURPOSES OF PATENT PROCEDURE

INTERNATIONAL FORM

RECEIPT IN THE CASE OF AN ORIGINAL DEPOSIT ISSUED PURSUANT TO RULE 7.3 AND VIABILITY STATEMENT ISSUED PURSUANT TO RULE 10.2

To: (Name and Address of Depositor or Attorney)

Brooks Air Force Base
Attention: Dr. Johnathan Kiel
8308 Hawks Road
Brooks AFB, TX 78235-5234

Deposited on Behalf of: Brooks Air Force Base

Identification Reference by Depositor:	ATCC Designation
Escherichia coli JM109, pIC20RNR, .	69905

The deposit was accompanied by: __ a scientific description _X_ a proposed taxonomic description indicated above.

The deposit was received September 27, 1995 by this International Depository Authority and has been accepted.

AT YOUR REQUEST:

_X_ We will inform you of requests for the strain for 30 years.

The strain will be made available if a patent office signatory to the Budapest Treaty certifies one's right to receive, or if a U.S. Patent is issued citing the strain and ATCC is instructed by the United States Patent & Trademark Office or the depositor to release said strain.

If the culture should die or be destroyed during the effective term of the deposit, it shall be your responsibility to replace it with living culture of the same.

The strain will be maintained for a period of at least 30 years after the date of deposit, and for a period of at least five years after the most recent request for a sample. The United States and many other countries are signatory to the Budapest Treaty.

The viability of the culture cited above was tested October 1, 1995. On that date, the culture was viable.

International Depository Authority: American Type Culture Collection, Rockville, Md. 20852 USA

Signature of person having authority to represent ATCC:

Frank Simione, Acting Director, Patent Depository

Date: October 3, 1995

cc: Charles Bricker

Fig. 7.10 Design of constructed plasmid that conveyed enhanced DALM biosynthesis in *E. coli* (plant nitrate reductase hybrid with endogenous nitrate reductase)

We found a mutant (Alls-Gifford strain; curlicue anthrax) that constitutively coiled on all media. It not only showed this phenotype but also grew slower, formed tiny colonies, and resisted the loss of its plasmids when grown at elevated temperatures, unlike all other strains of anthrax bacteria [8–12]. This phenotype was expressed naturally in *Bacillus anthracis* growing in the rhizosphere of plants [12, 13].

Attempts to biosynthesize DALM in other bacteria (various strains of *Escherichia coli*) led to early termination because of the lethality of the process, which revolves around the induction of the dissimilatory nitrate reductase (the conversion of the nitrate to nitrite in the 3AT medium is necessary to initiate 3AT polymerization) [14]. Anthrax bacteria are resistant to this lethal effect because of their nitrate reductase functions in the presence of oxygen, and the bacterium has other genes that convey resistance to lethality.

The anthrax nitrate reductase gene hybridizes with the plant assimilatory nitrate reductase. However, the nitrate reductases of other bacteria did not. The assimilatory nitrate reductase operates in oxygen environments. Dissimilatory nitrate reductases of bacteria only operate, without lethality, under microaerophilic or facultative anaerobic conditions or when nitrite is removed by being quickly reduced to nitrogen gas by other bacterial enzymes. Finally, an *E. coli* strain (JM109) and a plasmid (pIC2ORNR1.1) were designed that conveyed the expressible flavoprotein part of the barley plant nitrate reductase to the bacteria, which could then biosynthesize DALM (Figs. 7.12 and 7.13).

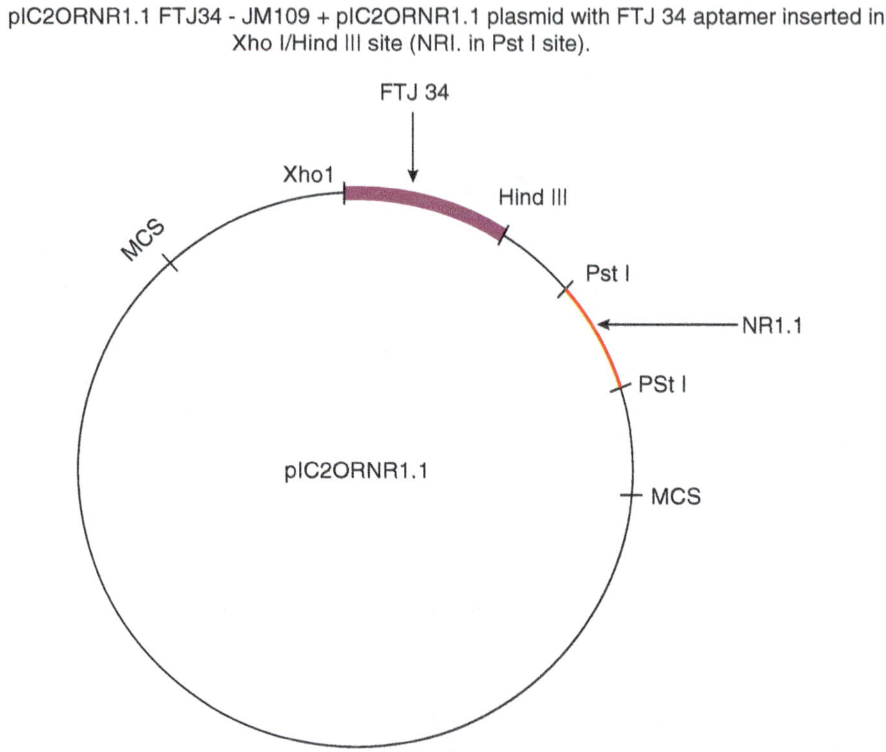

Fig. 7.11 The DALM-producing plasmid with an aptamer sequence inserted to make it more specific to attacking *Francisella tularensis*, the tularemia-causing bacteria. JM109 *E. coli* pIC2ORNR1.1 plasmid with FTJ 34 aptamer inserted in Xho I/Hind III restriction enzyme (endonuclease) sites (NRI. in Pst I site)

Fig. 7.12 Absorption spectrum of DALM dissolved in deionized water

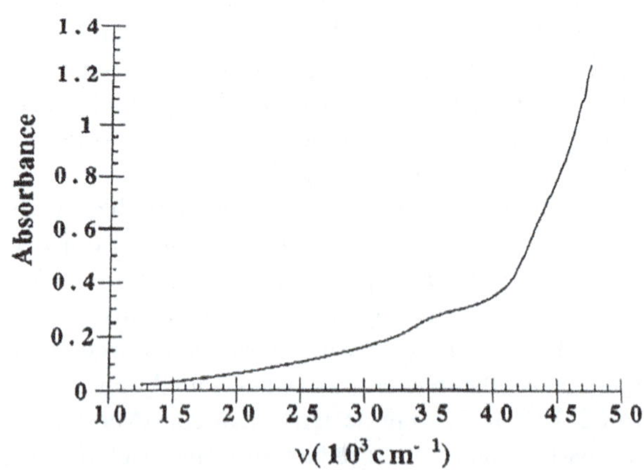

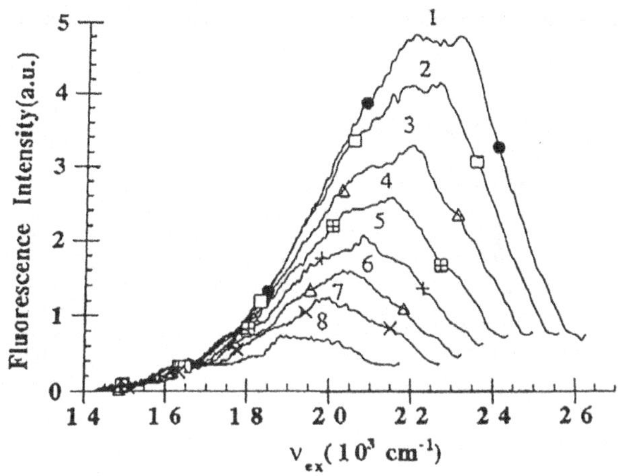

Fig. 7.13 DALM fluorescence emissions when activated at various excitation wavelengths. Emission spectra of DALM dissolved in deionized water as a function of near UV-visible excitation wavelength. Spectra are labeled according to the format spectrum wave number: excitation energy in wave numbers (excitation wavelength in nm): 1: 27,027 (370); 2: 26,316 (380); 3: 25,641 (390); 4: 25,000 (400); 5: 24,390 (410); 6: 23,810 (420), 7: 23,256 (430); 8: 22,222 (450)

After much time-consuming difficult analysis and modeling, the structure of DALM was finally realized. Initially, we thought its core structure was similar to that of the textbook structure of melanin (thus the name) and made simple empirical models and predicted absorption (and fluorescent) spectra, which in major absorption characteristics were similar to natural melanin.

DALM had a major absorption band at 365 nm and a calculated band at 323 nm (Table 7.1). Human melanin has an absorption at 335 nm (near the average of DALM actual and theoretical absorbances), with absorption almost completely attenuated for wavelengths longer than 700 nm. The other absorbances of DALM are "hidden," only revealed by corresponding fluorescence emissions (Figs. 7.14, 7.15, 7.16, 7.17, 7.18, 7.19, and 7.20).

The evaluation of the DALM structure moved from a polyhydroxyindole, melanin-like structure to diazo-linked rings (thus, the "diazo" in the name) to a direct hydroxy phenylene-linked polymer. The polymer was also unusual, being water soluble, unlike melanin, which exists as membrane-constrained granules composed of globular nanoparticles approximately 2 nm in diameter. DALM had an apparent pKa of approximately 5 by titration to precipitation, indicating the likely presence of carboxylic acid groups or more acidic hydroxyl groups. It formed insoluble salts with barium, calcium, and magnesium.

Even melanin did not support the textbook linear structure. Subsequent studies have shown it to be composed of circular oligomers, which imply higher molecular masses because of the spectra, but the latter being truly derived from the interactions of the stacked disks (intermolecular quenching of any emissions), not linear

Table 7.1 Observed and calculated optical absorbance and emission peaks (wavelengths) of diazoluminomelanin

Experimental		Calculated	
Absorbance	Emissions	Absorbance[a]	Emissions[b]
365 nm	480 nm[c]	323 nm	484 nm
	519 nm[d]		
	545 nm	446 nm	539 nm
	565 nm		565 nm
	578 nm	570 nm	581 nm
	590 nm		592 nm
	600 nm[e]		600 nm
	610 nm		606 nm
			610 nm
		695 nm	
		820 nm	
		1070 nm	
		1195 nm	
		1320 nm	

[a]Absorbance peaks calculated based on Kuhn's Theory of colored Conjugated carbon Compounds $\lambda_N = [1—(1/(2\,N + 1))]N \times 125.5$

[b]Emission peaks calculated based on $\lambda_d = 323[2—(1/d)]$, where d = the delocalization principal quantum number

[c]Major peak of fluorescence when excited with 365-nm light

[d]Major and

[e]principal minor peaks of chemiluminescence

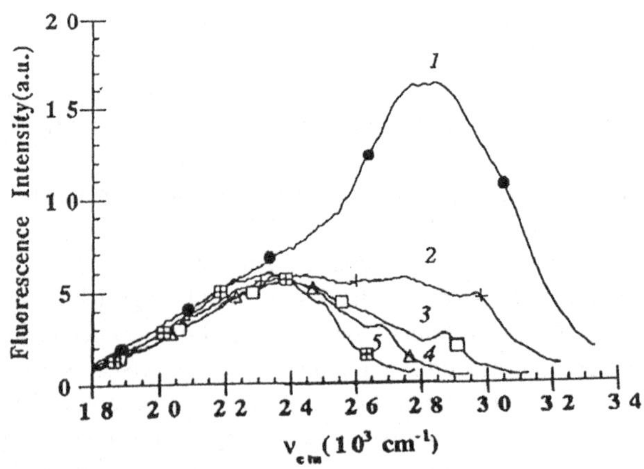

Fig. 7.14 Emission spectra of DALM dissolved in deionized water as a function of UV excitation wavelength. Spectra are labeled according to the format spectrum wave number: excitation energy in wave numbers (excitation wavelength in nm), 1: 34,483 (290); 2: 33,333 (300); 3: 32,258 (310); 4: 30,303 (330); 5: 28,571 (350)

polymerization (Fig. 7.21). DALM seems to be more like the textbook melanin structure than melanin.

Modeling the spectra of DALM from ab initio calculations led to two conclusions that explain much of its spectral and RFR and microwave behavior. First, its calculated spectra from this molecular modeling most closely resemble those of

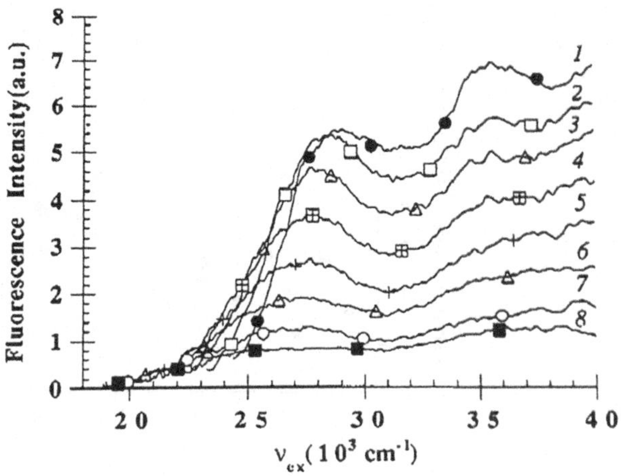

Fig. 7.15 Emission spectra of DALM dissolved in deionized water as a function of excitation wavelength. Spectra are labeled according to the format spectrum wave number: excitation energy in wave numbers (excitation wavelength in nm), 1: 23,810 (420); 2: 22,727 (440); 3: 21,739 (460); 4: 20,833 (480); 5: 20,000 (500); 6: 19,231(520); 7: 18,519 (540); 8: 17,857 (560)

Fig. 7.16 Excitation maximum as a function of emission wave number for DALM excitation spectra

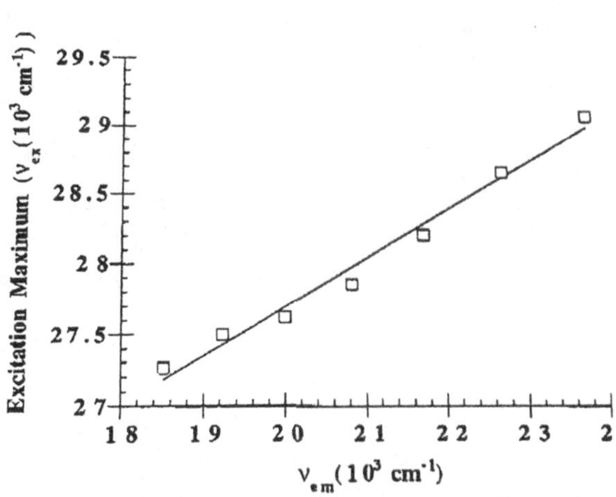

actual DALM polymers when they are linear hydroxy phenylenes and when the hydroxyls are ionized (negatively charged), and second, the lowest energy states are turned out of the plane with respect to each other (Fig. 7.20).

These models and observations suggest that the DALM traps the various states in “islands” of “planarity” between out-of-plane aromatic rings that can form quantum “boxes” of various sizes, thus absorbing and emitting different wavelengths based on the current energy state. The RFR and microwave pulses may correspond to certain quantized states of vibration and rotation that determine the sizes and lifetimes of the “boxes.” The E field was thought to drive the separation of charges along the ring and linear conjugated system of the polymer, making it a capacitor,

Fig. 7.17 Long lifetime component of fluorescence decay of DALM as a function of exciting wavelength

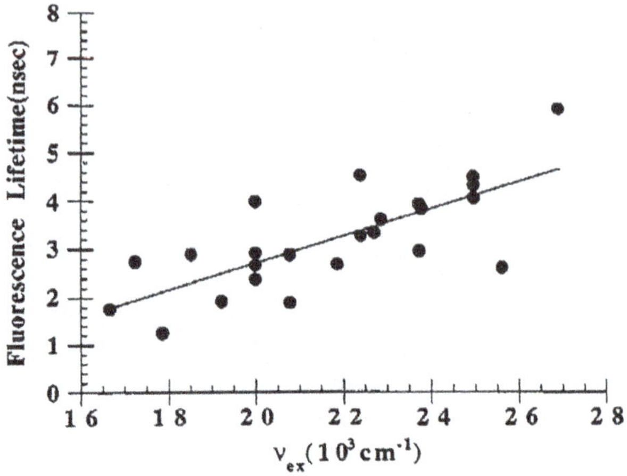

Fig. 7.18 Time-resolved emission spectra of DALM dissolved in deionized water as a function of excitation wave number (wavelength in nm). Spectra were collected 1 nanosecond after the peak of laser excitation. Spectra are normalized to their emission maxima

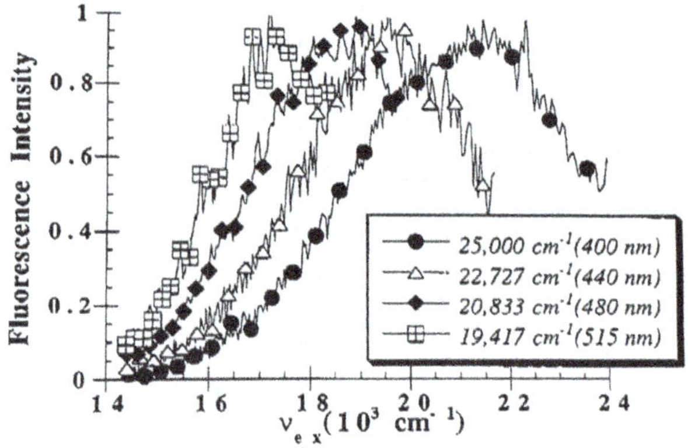

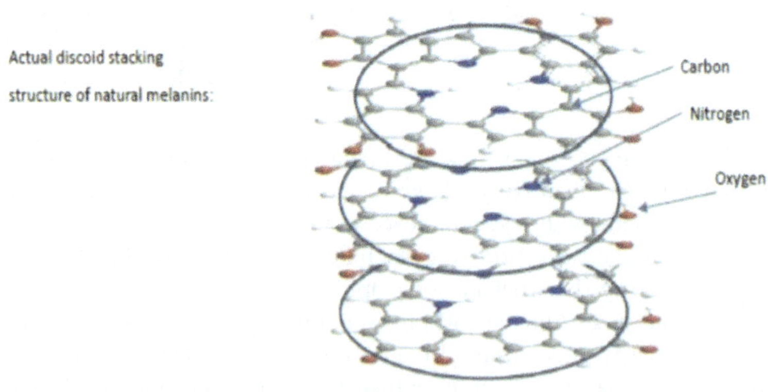

Fig. 7.19 The circular, stacked structure of natural melanin

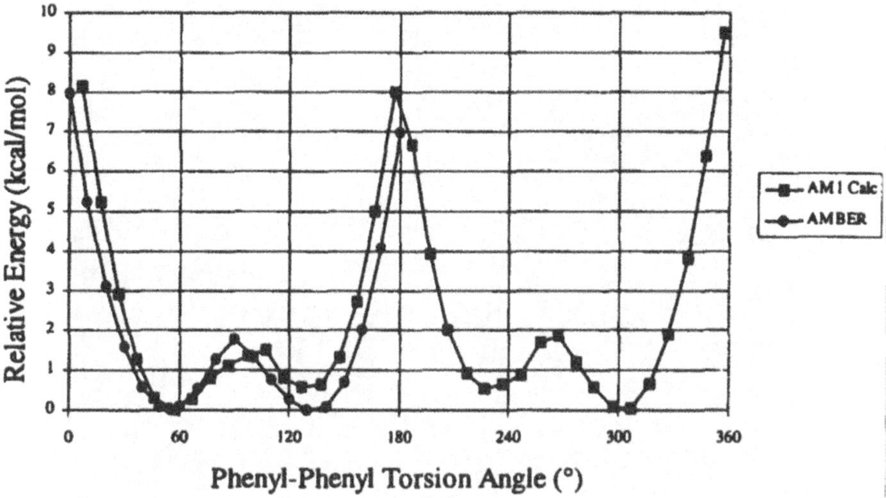

Fig. 7.20 The energy minima and maxima of the phenyl-phenyl relative angles with respect to the plane of the poly-phenylene DALM polymer. Heat of formation is shown for a biphenol system as a function of torsion angle using the AM1 Hamiltonian and AMBER force field methods, with a molecular dynamics simulation for a three-ring system

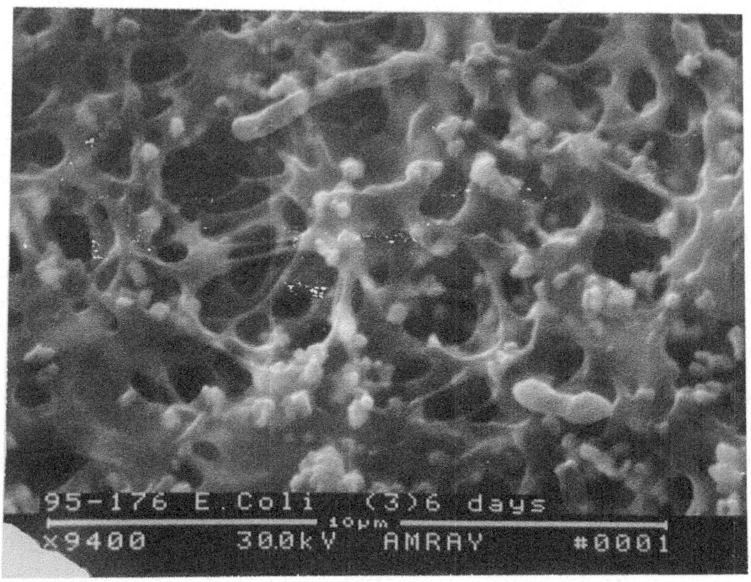

Fig. 7.21 Genetically modified *E. coli* on a filter after producing DALM (white globular material) to the point of breaking open the cells

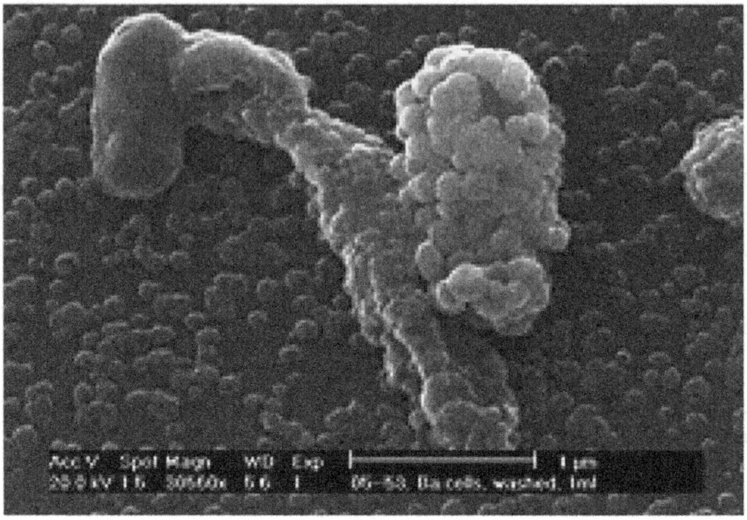

Fig. 7.22 Biosynthetic DALM/metallic nanoparticles on the surface of vegetative and spore forms of anthrax bacteria

but the dielectric properties showed that the polymer was actually acting as an inductor, storing the energy in a magnetic field. This observation means that the triplet (same spin electron pair) and doublet states (free radicals) are the dominant influencers of DALM's electromagnetic behavior. However, the separation of electron or ionic radical pairs does not preclude their entanglement and may lead to a "weird at distance" behavior, affecting whether they decay from triplet or singlet states (fluorescent electron pair of opposite spins) to ground state and whether they affect free radical interactions or simply release photons (slow fluorescence or thermochemiluminescence), regardless of how separated they are along the long axis of the polymer or whether they are in the same or distantly separated boxes. There may be a small measurement bias placed on the decay pathways due to this quantum effect: measuring thermalization or light emission affects the disentanglement pathway. The measurement affects the outcome. The final enigma to be resolved before moving on to applications and optimal designs for such application was the nanostructure association with the previously discussed responses.

Modeling suggested that we should find linear, fibrous, or ribbon structures that work best, but when the genetically modified organism (GMO) *E. coli*-biosynthesized DALM demonstrated properties equal to or better than the long ribbon synthetic form, our explanations were confounded (Figs. 7.21 and 7.22).

The scanning electron microscopy (SEM) of the *E. coli* after producing DALM illustrates three very important properties: (1) the DALM disrupts the bacterial cells, appearing to "dissolve" them away; (2) the DALM forms natural nanoparticles and nanoclusters; and (3) DALM nanoparticles and clusters are white in the EM, indicating a bright reflection of electrons, in turn, due to either the chelation of

Fig. 7.23 The separation of biosynthetic supercooled DALM solution liquid upon the freezing (−80 °C) of spent *E. coli* culture in a 3AT liquid medium

metal or, as a colleague of ours, Dr. John Wright, in these investigations proffered, the DALM acting like an "organic metal."

How can a linear polymer yield globular structures more like true melanin? We believe the answer lies in the ribbon molecular structure that wraps around a central core (could be bacterial protein, membrane, or cell wall subcomponents), like a ribbon ball. This conclusion was further supported by the observations that DALM readily and spontaneously coats iron nanoparticles and nickel/silver nanoparticles. Figure 7.22 shows such hybrid nanoparticles attached to anthrax spores and vegetative cells.

Such composite nanoparticles mediated rapid microwave killing of spores with a bound water coating but no additional activating redox agents. When dry, they were inert for all practical purposes. The interaction with water stabilizes the nanoparticles and facilitates the transfer of killing free radicals to microbial targets. This water interaction dramatically disrupted the structure of water. When placed in a −80 °C freezer, concentrated DALM was separated as a liquid supercooled supernatant from the frozen solid spent biosynthetic bacterial cultures (Fig. 7.23). This supercooled concentrated DALM solution would freeze when tapped but would return to a liquid state even well below freezing temperatures, which is characteristic of supercooled water. This observation provided the last theoretical piece to the complex behavior of DALM nanoparticles.

The states or phases of water in the supercooled liquid phase go beyond liquid water, vapor, and ice to two other liquid states of low and high density [15–17].

These latter two states are in equilibrium, especially at critical points (where the two liquid phases are indistinguishable), one observed and another theoretical. Pure water is limited to a supercooled lower temperature of −40 °C and, in certain plants that resist freezing, to −50 °C [18]. Our supercooled DALM solution was astounding, demonstrating supercooling at −80 °C. Such critical points for supercooled liquid-liquid phases have only been theoretically calculated and then calculated at very high pressures (1739 bars) [16]. Another property of supercooled water is that it can transfer phonons like a solid [19]. The low-density supercooled state provides voids, or defects, to collect gas to facilitate sonoluminescence. Thus, quantum phonon effects could lead to the amplification and concentration of sound (shock waves) to yield sonoluminescence and the audible sound corresponding to microwave pulses.

The finely divided metallic nanoparticles are temporarily protected against autoxidation (otherwise pyrophoric effects), but when they are activated by microwaves, they are reengaged in antimicrobial activity. These observations led to our first very important application of radio-controlled nanoparticle vectors.

References

1. Dehane, A., et al., A comprehensive numerical analysis of heat and mass transfer phenomenons during cavitation sono-process, Ultrasonics Sonochemistry 73: 105498, 2021.
2. Szeri, A., et al., Heat and mass transfer during the violent collapse of nonspherical bubbles, Phys. Fluids, 1(9), 2576–2586, September 2003.
3. Dehane, A., et al., Insight into the impact of excluding mass transport, heat exchange and chemical reactions heat on the sonochemical bubble yield: Bubble size-dependency, Ultrasonics Sonochemistry 73: 105511, 2021.
4. Yuan, G., et al., Two Mechanisms Determine Quantum Dot Blinking, https://chemrxiv.org/engage/api-gateway/chemrxiv/assets/orp/resource/item/60c73d0dbdbb894e96a37bed/original/two-mechanisms-determine-quantum-dot-blinking.pdf.
5. Pelton, M., et al., Evidence for a diffusion-controlled mechanism for fluorescence blinking of colloidal quantum dots, PNAS 104 (36): 14249–14245, Sept 4, 2007.
6. Auger Recombination, LibreText, https://batch.libretexts.org/print/url=https://eng.libretexts.org/Bookshelves/Materials_Science/Supplemental_Modules_(Materials_Science)/Solar_Basics/C._Semiconductors_and_Solar_Interactions/IV._Recombination_of_Charge_Carriers/3._Auger_Recombination.pdf.
7. Kiel, J., Parker, J., Alls, J., and Weber, R., Self-Labeling of Bacteria with a Luminescent Polymer. In Proceedings of the Thirteenth Annual Conference of the IEEE Engineering in Medicine and Biology Society, (J.H. Nagel and W. M. Smith, eds.), Vol. 13, No. 4, IEEE, Philadelphia, pp. 1605–1606 (1991).
8. Kiel, J., Alls, J., Weber, R., and Parker, J., Rapid Identification of Environmental Bacillus. U.S. Patent 5,156,971 (issued 20 Oct 92).
9. Bruno, J.G., and Kiel, J.L. Effect of Radio-frequency Radiation (RFR) and Diazoluminomelanin (DALM) on the Growth Potential of Bacilli. In Electricity and Magnetism in Biology and Medicine, (Blank, M., ed.), San Francisco Press, San Francisco, pp. 231–233, 1993.
10. Parker, J., and Kiel, J., US Patent 6,878,519 B1. DNA Probes for Specific Genes of Anthrax, 12 April 2005.
11. Parker, J., Kiel, J., Gifford, H., Alls, J., and Morales, P., U.S. Patent 7,279,320 B1. Curlicue Vaccine Strain of Bacillus Anthracis, 9 Oct 2007.

12. Kiel, J., Parker, J., Alls, J., Kalns, J., Holwitt, E., Stribling, L., Morales, P., and Bruno, J., Rapid recovery and identification of anthrax from the environment. In Tropical Veterinary Diseases, J. House, K.M Kocan, and P. Gibbs, Eds., Annals of the New York Academy of Sciences 916: 240–252, Sept 2000
13. Kiel, J., Parker, J., Grubbs, T., and Alls, J., Growth medium for the rapid isolation and identification of anthrax. In Chemical and Biological Sensing, P. J. Gardner, Ed., Proceedings of SPIE 4036: 92–102, 2000.
14. Kiel, J.L., Parker, J.E., Holwitt, E.A., and Schwertner, H.A. U.S. Patent 5,856,108 (AF Inv 19,827) Biosynthesis of Diazomelanin and Diazoluminomelanin, 5 Jan 1999.
15. Chaplin, M., Water Structure and Science. Water.isbu.ac.uk. 25 September 2022.
16. Debenedetti, P.G., Sciortino, F., and Zerze, G.H., Second critical point in two realistic models of water, Science 369: 289–292, 2020.
17. Pihlajamaa, I, Laudicina, C.C.L., Luo, C., and Janssen, L.M.C., Emergent structural correlations in dense liquids, PNAS Nexus 2 (6): 1–9, 2023 (https://doi.org/10.1093/pnasnexus/pgad184).
18. Cavender-Bares, J., 19-Impacts of freezing on long distance transport in plants, Vascular Transport in Plants, Physiological Ecology, pages 401–424, 2005.
19. Frank, M., and Drikakis, D., Solid-like heat transfer in confined liquids, Microfluid Nanofluid 21: 148–154, 2017.

Chapter 8
Nonionizing Electromagnetic Radiation-Driven Microbiological Effects

Abstract Even to date, there has been a simplistic search, heretofore unsuccessful, for a frequency of radiofrequency radiation (RFR) or microwaves that may specifically kill pathogenic microbes selectively. As we have already shown, any such specificity should be guided by a directed sensitizer that engages specific frequency ranges, with correlated penetration depths, into the microbial matrix and appropriate intensities to minimize collateral damage to materials or tissues. Detection, identification, and monitoring of microbial contact and microbial neutralization are the goals of any such radio-controlled vector. Furthermore, these vectors must penetrate microbial cells or, in the case of viruses and intracellular microbes, host cells and genetically alter the microbes' susceptibility to antibiotics, other antimicrobial agents, and neutralization by the immune system. This chapter discusses what has been successful, to date, toward these goals; the pathogens attacked; and the potential for countering other pathogens in the environment and in hosts.

Keywords Aptamers · Anthrax · *Rickettsia* · Tularemia · Plasmids · Quenching · Quantum dots · Luminol

Our search for the effects of radiofrequency radiation (RFR) on microbes began with the development of the process for the biosynthesis of diazoluminomelanin (DALM) in various strains of *Escherichia coli,* which started soon after its first successful chemical synthesis on March 1, 1988, and in 1989, it was extended to genetically modified strains to optimize production. The need for biosynthesis was primarily to avoid the use of organic solvents such as DMSO (dimethyl sulfoxide) and acetone and to enhance scale-up in production. These solvents were environmentally problematic in their disposal, although we recycled them as much as possible (redistilling acetone) (Table 8.1).

By the end of 1990, we had developed the 3AT bacterial growth medium into one that would selectively grow *Bacillus anthracis*, accelerating its growth so that

---

The original version of the chapter has been revised. A correction to this chapter can be found at https://doi.org/10.1007/978-3-031-63006-4_12

J. L. Kiel, *Bionanotechnology*, https://doi.org/10.1007/978-3-031-63006-4_8

Table 8.1 Slow luminescence and nitrate reduction and nitrogen production of bacteria grown on various substrates (Parker et al., US Patent 5,902,728, May 11, 1999). A comparison of various species and strains of *Bacillus* and *E. coli* (including one containing the genetically modified plasmid for enhanced DALM production) for DALM production using various potential precursors (3AT = 3-amino-L-tryrosine; 3-NitroT =3-nitrotyrosine; 2,4-DCP = dichlorophenol. NR = nitrate reductase activity) [1]

Slow luminescense of nitrogen-metabolizing bacteria						
	Medium					
Species	TSA	2X3AT	3-NitroT	2,4-DCP	NR	N_2
Bacillus anthraces	20.3	39.0	22.1	13.3	+	–
Bacillus cereus	52.6	184.6	25.7	33.6	+	–
E.coli JM109/	46.1	860.0	142.1	58.9	+	–
pIC20RNR1.1						
E.coli JM109	75.0	199.0	12.9	61.6	+	–

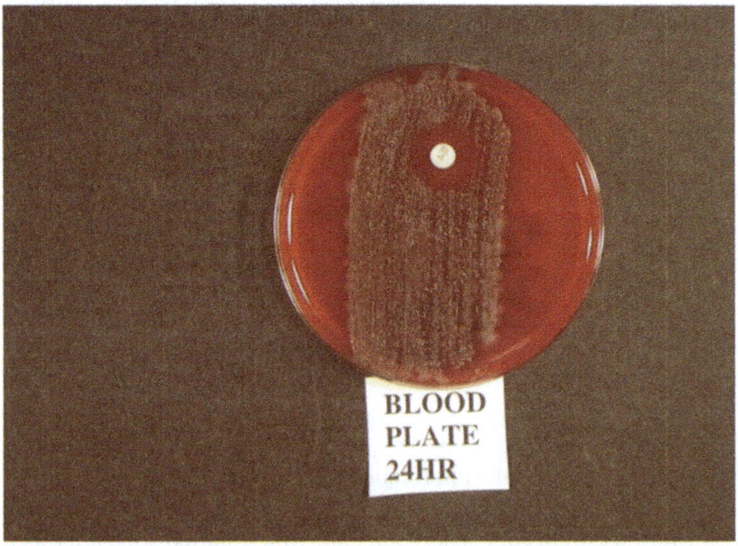

Fig. 8.1 The interference with tests for anthrax bacteria by environmental bacterial contamination

identifying tests could be completed in 4–8 h rather than 24–48 h and sensitizing it to nonionizing electromagnetic radiation (NEMR) killing to avoid further contamination [2]. Figures 8.1 and 8.2 show a comparison of environmentally contaminated *Bacillus anthracis* grown on blood agar versus 3AT medium to which gamma bacteriophage and penicillin sensitivity tests have been applied.

Contaminating bacteria obscured the tests on blood agar by overgrowth, but the 3AT medium suppressed the other bacteria sufficiently to discern the characteristic reactions of anthrax bacteria. The 3AT medium first accelerated growth in the first few hours, then led to the inhibition and death of the bacteria, even *Bacillus anthracis* (Figs. 8.3 and 8.4).

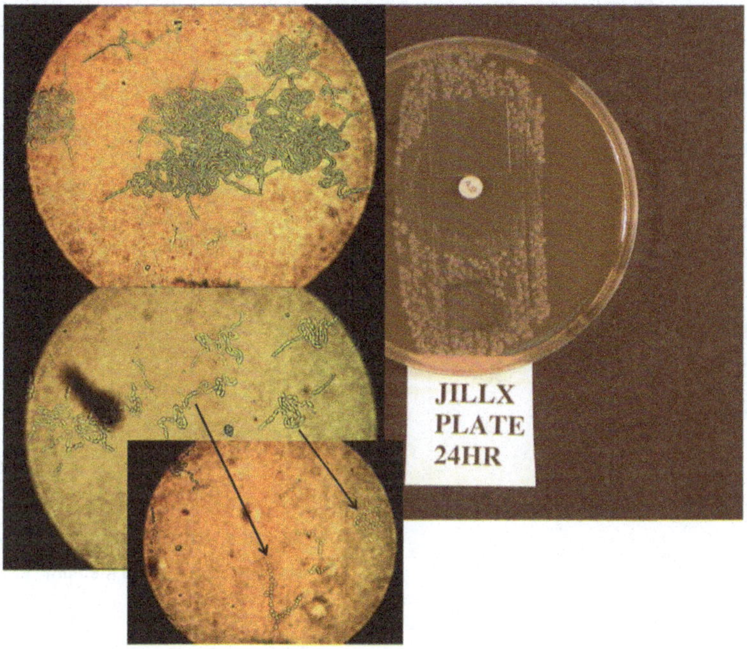

Fig. 8.2 Anthrax tests (Cherry bacteriophage and penicillin string of pearls) on 3AT medium revealed the suppression of environmental contaminating bacteria. Microscopy inset pictures showing a string of pearl reaction and Medusa head growth

The photomicrographs (Fig. 8.2) show the coiling phenotype induced by the 3AT medium and the rapid "fixed" "string of pearls" reaction to penicillin (or ampicillin).

The 3AT medium stops the pearls from bursting osmotically and sustains the structure from the loss of cell wall. Penicillin causes this fatal loss of the cell wall by its inhibitory binding to the beta-lactam ring of DD-transpeptidase, which otherwise cross-links a chain in the cell peptidoglycan, maintaining the integrity of the cell wall in a low osmotic environment. At first, we believed that this fixation and the accompanying coiling were a result of polymer formation linked to cell wall elements. However, the *E. coli* formation of globular nanoparticles contradicted this assumption.

The next evidence, for which some other explanation applied, was the isolation of a pinpoint colony mutant of *Bacillus anthracis* that constitutively coiled, even without the 3AT medium components. This mutant did not lose the anthrax plasmids when grown at elevated temperatures, unlike other strains of *Bacillus anthracis* [3]. The photomicrographs in Figs. 8.5, 8.6, 8.7, 8.8, and 8.9 show these properties in various growth media.

Another incorrect assumption was that DALM could not penetrate bacterial or other cells because of its molecular size and negative charge. This incorrect conclusion was reenforced by clawed frog teratogenicity assays showing that no teratogenicity occurred with DALM treatment [4]. DALM was assumed not to cause teratogenic toxicity because it could not penetrate the test cells.

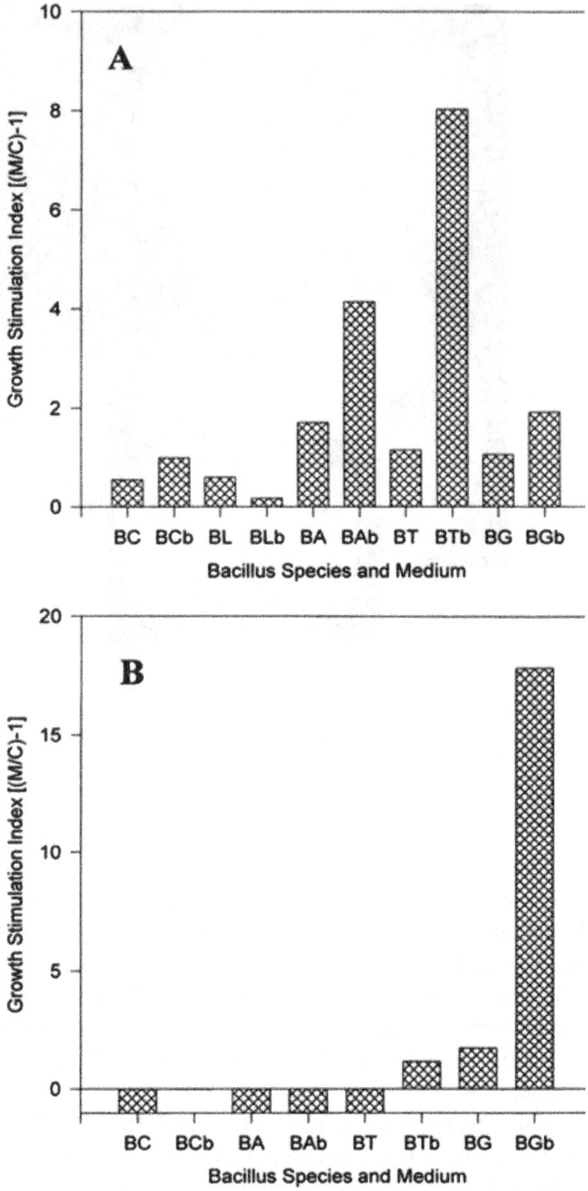

Fig. 8.3 The 3AT medium first accelerated growth in the first 6 h (**a**) and then led to the inhibition and death (at 24 h, **b**) of *Bacillus globigii v. niger* (BG), *B. licheniformis* (BL), *B. anthracis* (Sterne strain; BA), *B. cereus* (BC), and *B. thuringiensis v. kurstaki* (BT), especially when sodium bicarbonate was included (b). 3AT medium with (b) and without sodium bicarbonate added

The search for new methods in biological agent counterproliferation (supplanting the work on the personal RFR dosimeter) led to more compact, immediate, and user-friendly detection, isolation, and identification methods that safely contained a viable pathogen for further analysis and forensics. Viability assessment was very important, as the earlier but less contained methods illustrated. Because of the assumptions that DALM could not penetrate bacterial cells but could be produced

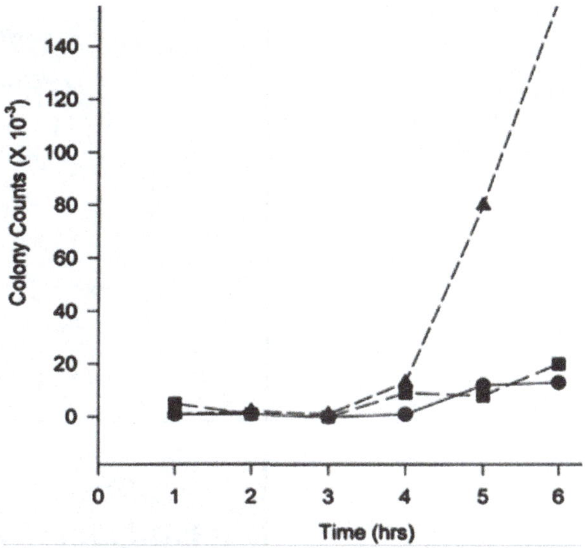

Fig. 8.4 The growth kinetics of *Bacillus anthracis* in TSB (closed circles), 3AT liquid medium (closed squares), and 3AT liquid medium with bicarbonate (2 g/L, closed triangles). The pH of TSB was 7.3, 3AT was 7.26, and 3AT with bicarbonate was 7.7

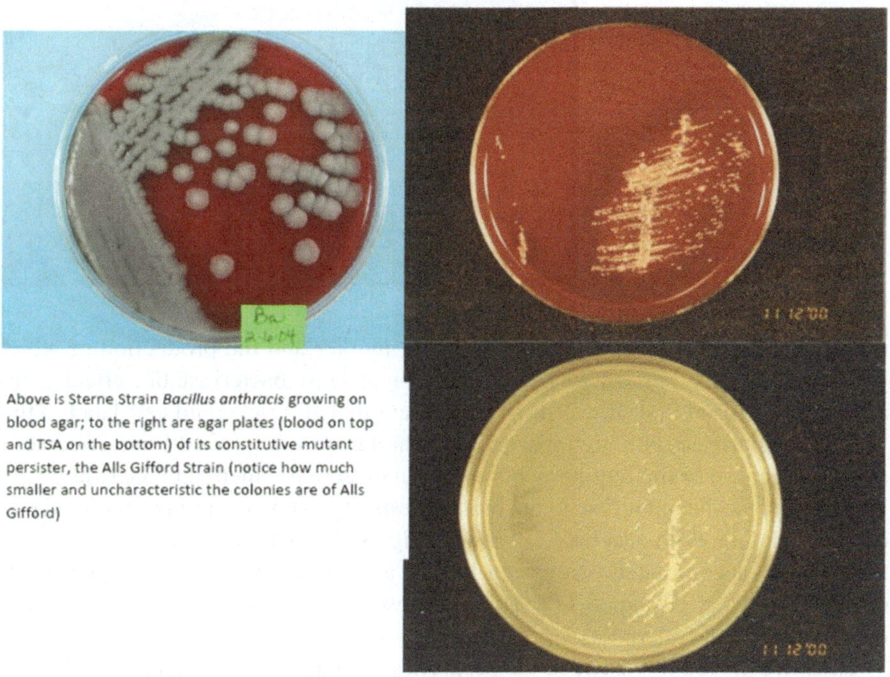

Fig. 8.5 Alls-Gifford (Curlicue) strain of *Bacillus anthracis* (on the right) growing on blood and TSA agars, demonstrating its pinpoint colonies due to constitutive coiling compared to typical anthrax bacterial growth on blood agar (on the left)

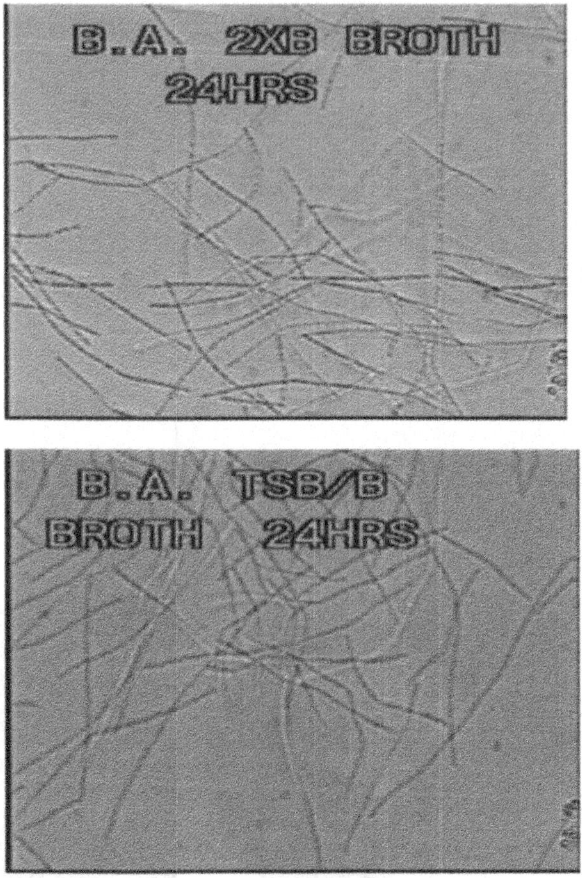

Fig. 8.6 Growth of Sterne strain (top pair) anthrax bacteria grown in TSB and 3AT liquid media, respectively, for 24 h showing growth characteristics

in situ as a useful fluorescent or luminescent marker and the production of which indicated viability but led to bactericidal, or at least bacteriostatic, effects, that approach was used first. It was most convenient that anthrax could self-label in this way. This approach, although it worked in the laboratory, proved impracticable in the field because of the requirements for specialized equipment and analysis.

The fluorescence first rose in the biosynthesis of DALM bacteria (*E. coli* JM109/pIC2ORNR1.1), with accelerated growth (Fig. 8.10), but then it auto-quenched as the polymer accumulated or grew in length, and the growth rate correspondingly declined (Figs. 8.10 and 8.11). The luminescence required redox reagents and energy to be manifested.

Next, we moved to the use of the properties of metal chelation and laser-induced line spectroscopy [5]. This approach worked well for tagging bacteria but was limited in identifying and testing for viability. Additionally, it requires specialized equipment and skills.

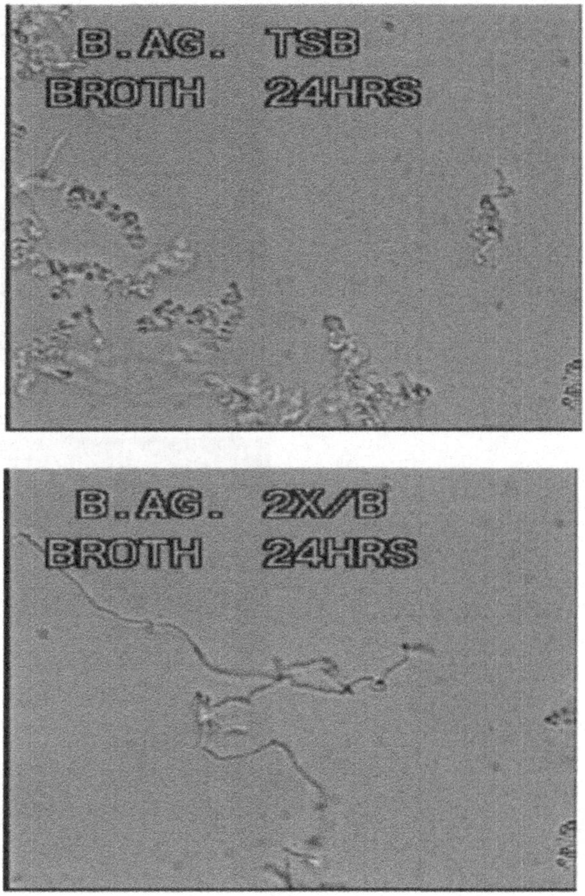

Fig. 8.7 Growth of Alls-Gifford strain (bottom pair) anthrax bacteria grown in TSB and 3AT liquid media, respectively, for 24 h showing growth characteristics

We used such metal tagging of bacterial spore surrogates for anthrax in full-scale tests using rare-earth scandium. Where viability was lost, it could still indicate the distribution of spore remnants.

Next, we decided to take advantage of DALM's powerful quenching ability and the strong, stable fluorescence of quantum dots. In this case, rather than the brute force technique of rare earth metal tagging, we used quantum dots quenched with a black hole quencher or DALM. We used double-stranded DNA aptamers made to bind specifically to target pathogens or toxins that dequenched when the target bound to the specific strand containing the quantum dot or other fluorescent marker and released the complementary one with the quencher (Figs. 8.12 and 8.13) [6–9].

This technology was extended to field trials with quenched dots applied to the wings of live flies that were released in the vicinity of released surrogate *Bacillus thuringiensis* (Fig. 8.14).

The flies were trapped, and those with fluorescent dots on their wings were cultured for Bt. The recovery rate was approximately 2% of the tagged flies released.

Fig. 8.8 Scanning electron of the Alls-Gifford strain of anthrax bacteria

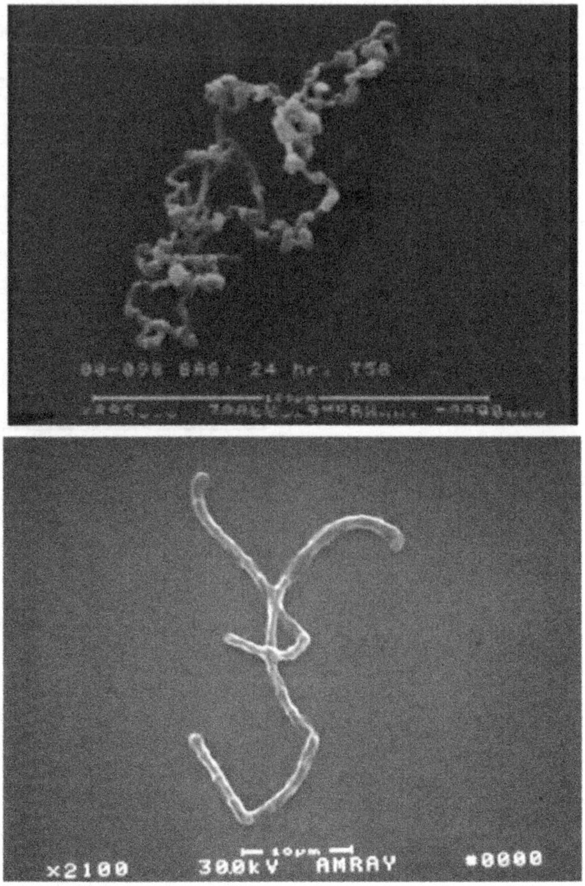

This technology could be used as such to gather field samples in environmental niches entered by the respective insects or be further applied to detection, tentative identification, and safe collection field devices (Figs. 8.15, 8.16, 8.17, and 8.18).

Furthermore, the detected and collected pathogen could be safely returned to the rear echelon, confined in a cassette, which could be disinfected on the outside but still contain viable organisms, for further culture or genetic analysis, such as polymerase chain reaction (PCR) or sequencing (Fig. 8.17).

This system was validated with field samples but never deployed.

The next step was to tie detection and identification to killing the pathogen target. Here, we revisited DALM killing activated by chemical reactants and RFR or microwave radiation. The problem was the chemical activation part. There were two steps in solving this problem. We had already demonstrated that we could tag and confirm the tagging of a pathogen in a complex environment and recover it. The next step was to find an activation methodology that could be applied to killing

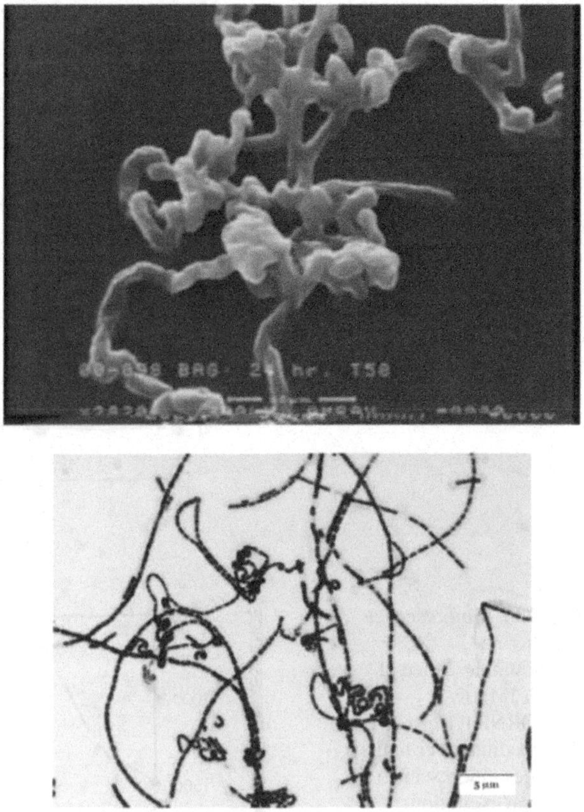

Fig. 8.9 Scanning electron and light microscopy (Gram stained) of the Alls-Gifford strain of anthrax bacteria

bioagents and be stopped at will to prevent collateral damage to sensitive materials, people, and other living entities. Inherent to this solution, the DALM nanoparticles would have to be easily dispersed over a large area that could be addressed with a wide-area activation process.

We first examined another neutralization technology, cold plasma. This technology generates ionized gases, added as mixtures, or activated atmospheric gases to destroy environmental pathogens. This technology and its accompanying variety of devices tested held out the promise of in-field generation of disinfecting gases with no persistent collaterally damaging chemicals [10, 11].

The question was whether these activated gases activate the killing potential of DALM nanoparticles bound to bioagents. The alternative was activating the DALM nanoparticles by the application of RFR or microwave radiation and ceasing killing on the termination of NEMR application. The initial step was to analyze the killing of existing cold plasma devices and the effective gases they produced. The first was a pulsed corona plasma reactor pictured in Fig. 8.18 and modified for our laboratory to collect gases at various points along a flow path where bacterial spore coupons were placed to correlate gas analysis with killing potential without DALM.

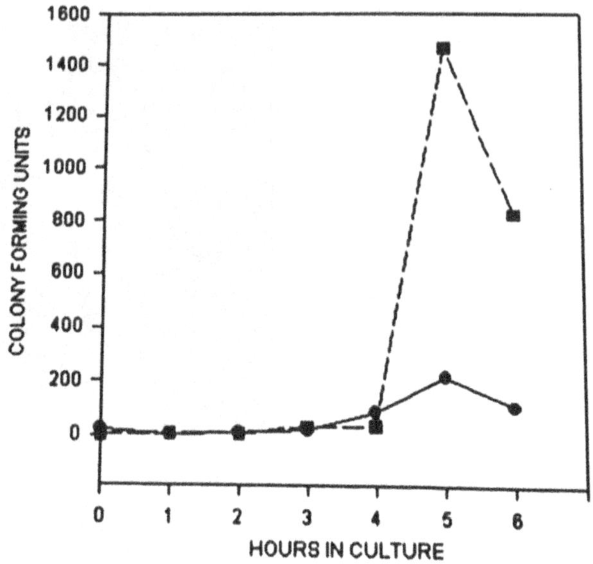

Fig. 8.10 *E. coli* JM109/pIC2ORNR1.1 growth kinetics on 3AT liquid medium (closed squares) vs TSB (closed circles)

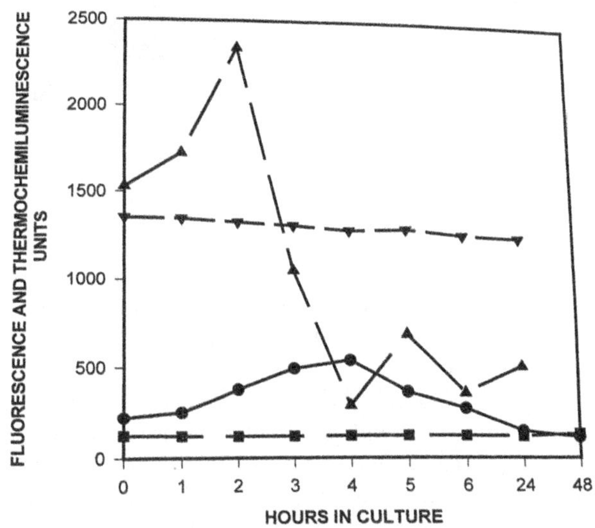

Fig. 8.11 Fluorescence and thermo-chemiluminescence of *E. coli* JM109/pIC2ORNR1.1 grown on 3AT medium over hours; triangles, fluorescence (365 nm excitation, 450 nm emission) of the cultures; inverted triangles, fluorescence of medium without bacteria; closed circles, thermo-chemiluminescence of culture (10-s integrations of each hourly sample); closed squares, medium (control) thermo-chemiluminescence

In summary, this device killed further out from the corona discharge than expected. It used air and no additional supplemental gases. The active agent proved to be singlet oxygen.

Other potential agents, such as ozone and nitrogen oxides, proved not to be the persistent killing agent generated. Hydroxyl radicals, another possible killing agent, which can persist when diluted in air, are so reactive with many other materials that they do not transit materials but react mostly at the surface. Additionally, ozone usually requires water or a sensitizing agent (yielding organic peroxides) to effect

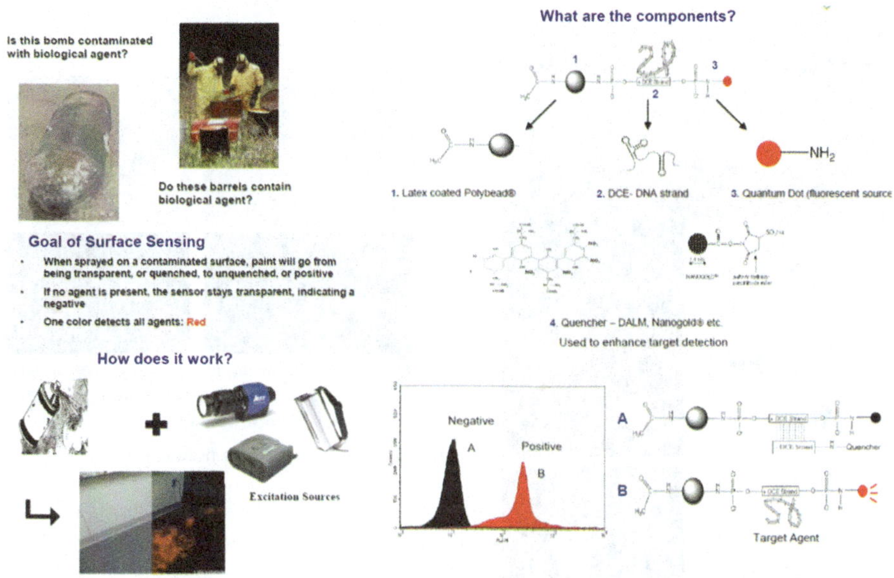

Fig. 8.12 Scheme of surface detection of biological warfare agents on contact with quenched quantum dots or fluorophores bound to double-stranded DNA aptamers that dequench when they bind to agents

killing. They also constitute a persistent toxic gas hazard to human health and are environmentally damaging pollutants. Singlet oxygen is less of a toxic gas hazard because of its lack of persistence and its photochemical decay [12–14]. It can be derived from monatomic oxygen first formed from ozone interacting with ultraviolet (UV) light and then interacting with a second ozone molecule to produce singlet oxygen, the destruction of ozone in its formation being an additional benefit. However, it still requires interaction with organic or biological targets (such as unsaturated lipids to form organic peroxides) to convert it to an imminent bactericidal or bacteriostatic agent.

The next set of devices to be examined were those that used various mixtures of inert gases, such as argon, and active oxygen to form optimal cold plasma gases. The greater effectiveness of more dilute mixtures of oxygen suggests that a bimolecular oxygen reaction occurs that dissipates the energy necessary for the killing mechanism. Even though two molecules of delta oxygen yield a higher-energy sigma oxygen, which would be expected to be more reactive, the bimolecular reaction, which removes half the singlet oxygen and leaves higher-energy singlet oxygen, is expected to have a shorter lifetime and shorter reactive reach (Fig. 8.19).

With the devices examined, we obtained quantitative results with *Bacillus thuringiensis* (Bt) spores (Figs. 8.20, 8.21, 8.22, 8.23, and 8.24).

We modified one device to control the distance and time of application of the cold plasma to spore-coated coupons to more precisely determine optimal killing conditions. We showed that cold plasma was a very efficient, low-residue killing

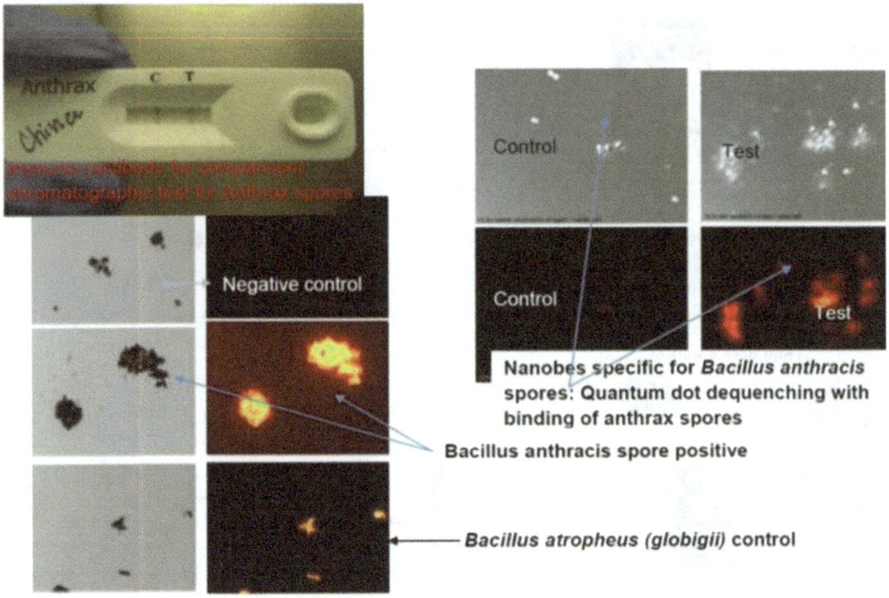

Fig. 8.13 Comparison of a lateral flow antigen diagnostic assay for anthrax with an aptamer dequenching assay (at the microscopic level)

Fig. 8.14 Field tests in Honduras of insect capture, label with quenched specific detection aptamers, release and recapture fly borne detection of bioagents

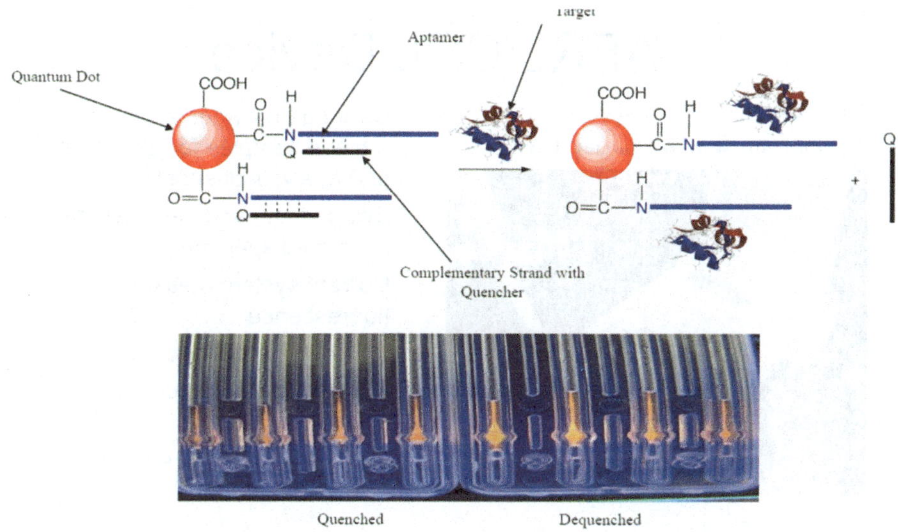

Fig. 8.15 Diagram of the double-stranded DNA aptamer dequenching assay and actual micro-channel dequenching assays

Fig. 8.16 Magnetic capture and dequenching chamber for the safe detection and collection of especially dangerous pathogens

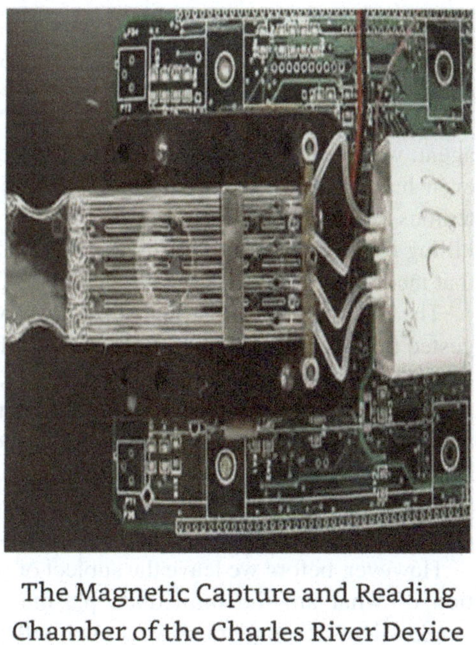

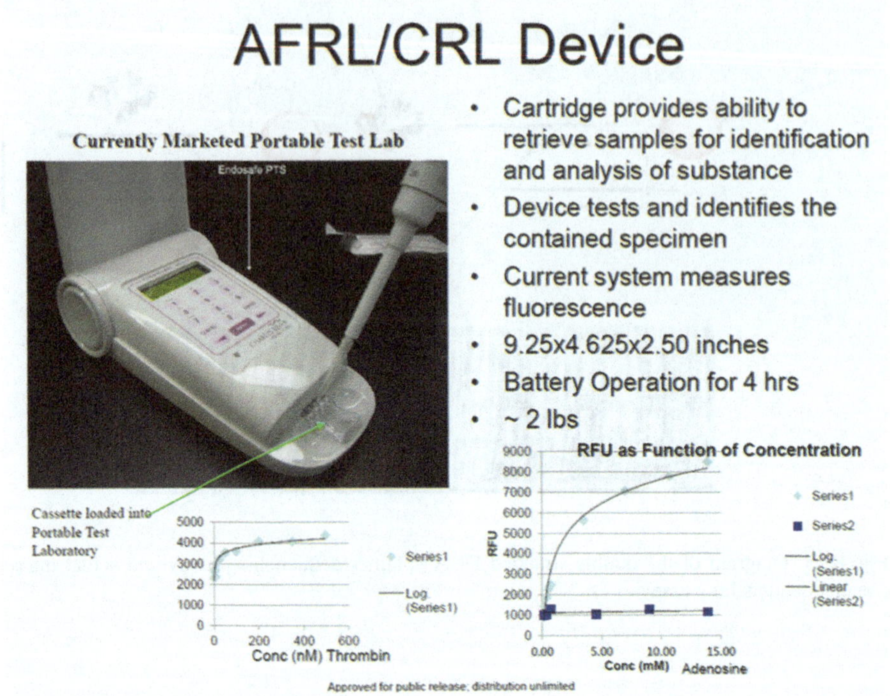

Fig. 8.17 The completed device for magnetic capture and dequenching aptamer detection and identification of pathogens and toxins

agent, with little damage to sensitive devices (such as computers, cell phones, and even human skin) under optimal killing conditions. However, its application is tedious and time-consuming, especially for large surface areas. The quenching of killing by DALM strongly suggested that DALM reacted with the activated gases that mediated killing (Figs. 8.22, 8.23, 8.24, and 8.25).

The flashing and electrical discharges with pulsed microwaves strongly suggested that the immediate killing by DALM nanoparticles was indeed mediated by its generation of plasma gases, nanoplasma, on the immediate surfaces of bacterial targets. The indicator technology, the dequenching of fluorescence upon specific agent binding, could be used to monitor agent destruction on surfaces in real time by its disappearance with plasma treatment. The plasma destroys not only chemical fluorophores but also the fluorescence of quantum dots that correlate with the bioagent's destruction. The two technologies were now joined.

However, before we leave the subject of plasma kill, we need to address the question of what are the immediate plasma gas components generated by DALM nanoparticles through microwaves that kill microbes? The reactant requirements for an optimal response to microwaves give us a clue. The reaction mixtures contain DALM, hydrogen peroxide (at low concentrations to initiate the reactions but not

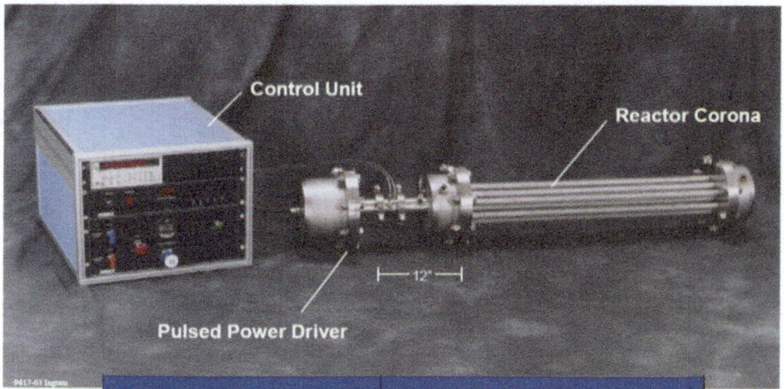

Gas Flow Rate	Depends on Desired Destruction
Average Operating Power	Variable to 1 kW
Maximum Pulse Voltage	Variable from 10 kV to 30 kV
Pulse Repetition Rate	Variable to 2000 Hz

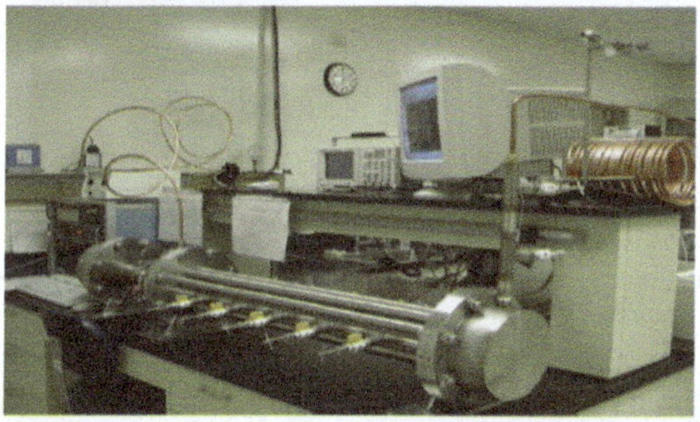

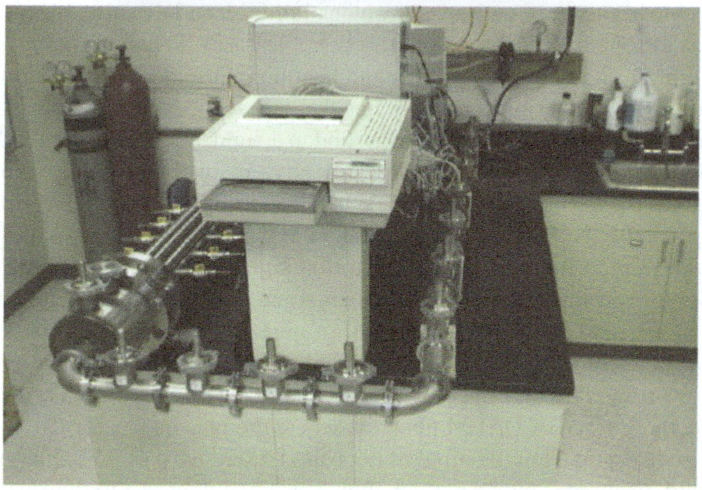

Fig. 8.18 Pulsed corona discharge device that used ambient air to generate cold plasma, with additional tubing added downstream from the reactor with ports for exposing bioagents and collecting gas samples

Fig. 8.19 Chemical reactions forming and converting singlet oxygen

$$O_3 + h\nu \rightarrow O_2 + O$$

$$O + O_3 \rightarrow O_2{}^1\Delta + O_2$$

$$O_2{}^1\Delta + O_3 \rightarrow 2O_2{}^3\Sigma + O$$

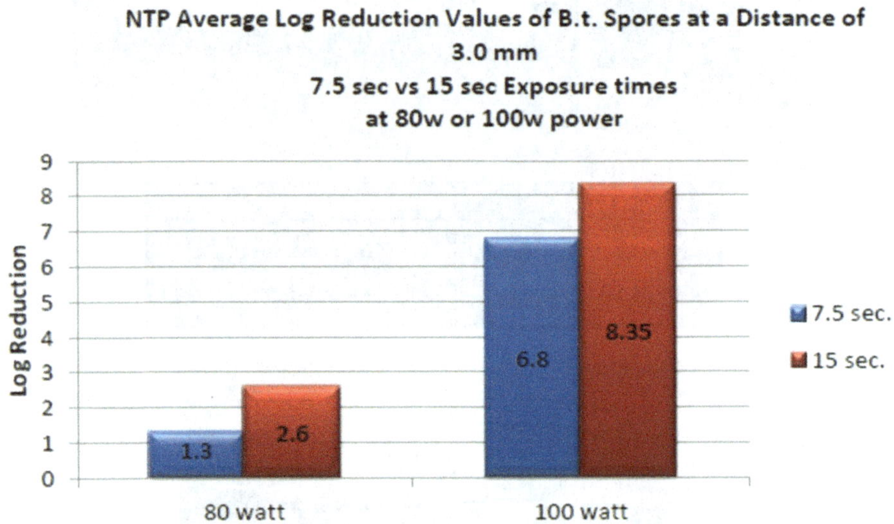

Fig. 8.20 Log reductions (tenfold iterations of killing) of *Bacillus thuringiensis* spores with various cold plasma exposures

enough to maintain them for long periods), a carbon dioxide source (such as sodium carbonate, bicarbonate, or carbonated water), luminol, and dissolved gases from the air, such as nitrogen and oxygen and rarely noble gases. When luminol oxidizes to produce light, it releases nitrogen gas, and heating facilitates the breakdown of carbonates into carbon dioxide and facilitates the degassing of the aqueous solution.

A casual observation of the reaction under a light microscope provided some very important data. When the DALM fibers were in the reacting solution, gas emerged from the surface of the fibers. When these were in the brightest part of the microscopic field, the bubbles emerged faster and pushed the fibers out of the field into a darker region where the gas generation significantly slowed. Each time the fibers were moved into the light, the gas emerged faster and pushed them into the dark part of the field, acting as a source of propulsion. This response indicated that the rate of the chemical reaction was controlled to a significant extent by a photochemical process. This interface of DALM polymer, liquid, and gas (the latter as bubbles on the surface of DALM fibers) is where the plasma gas forms, the electron temperature heats up from the input of deposited microwave or RFR energy, and the electric field discharges either in the bubbles or when they rise to the surface of the liquid and the gas is released into the air above the liquid. All this would account for the observations, the high temperatures within the bubbles and the rapid hot electron and free radical damage to the spores (the "popcorn effect").

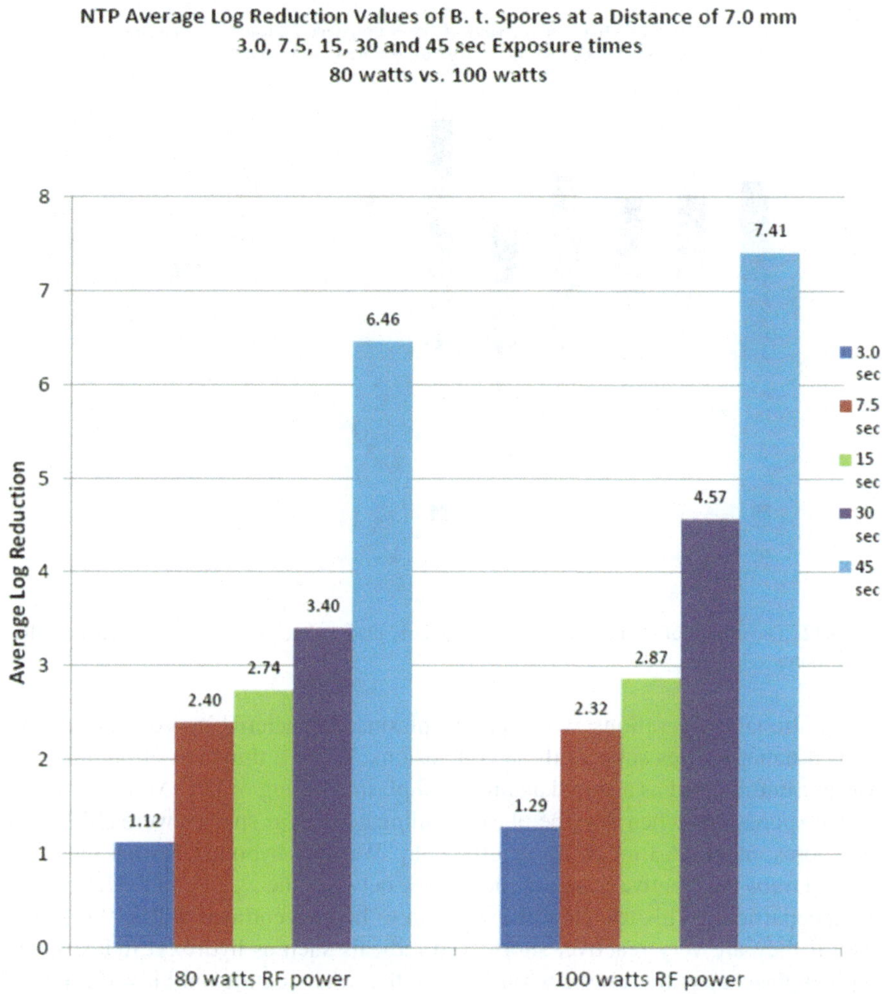

Fig. 8.21 Log reductions (tenfold iterations of killing) of *Bacillus thuringiensis* spores with various cold plasma exposures

This set of observations and processes applied to the DALM high-aspect ratio ribbons and fibers must also apply to the spherical DALM nanoparticles, especially those with metallic cores, because of the concentration of the electric fields at the surface of the metallic particles within the DALM semiconductor (leading to a circular current). Although DALM, unlike melanin and porphyrins (in heme proteins and cytochromes), is not circular, in the ribbon structures wrapped around the metallic nanoparticle cores, they become "ribbon balls," accomplishing the same thing. Scanning electron micrographs revealed particles and clusters of particles close and in association with bacterial spores.

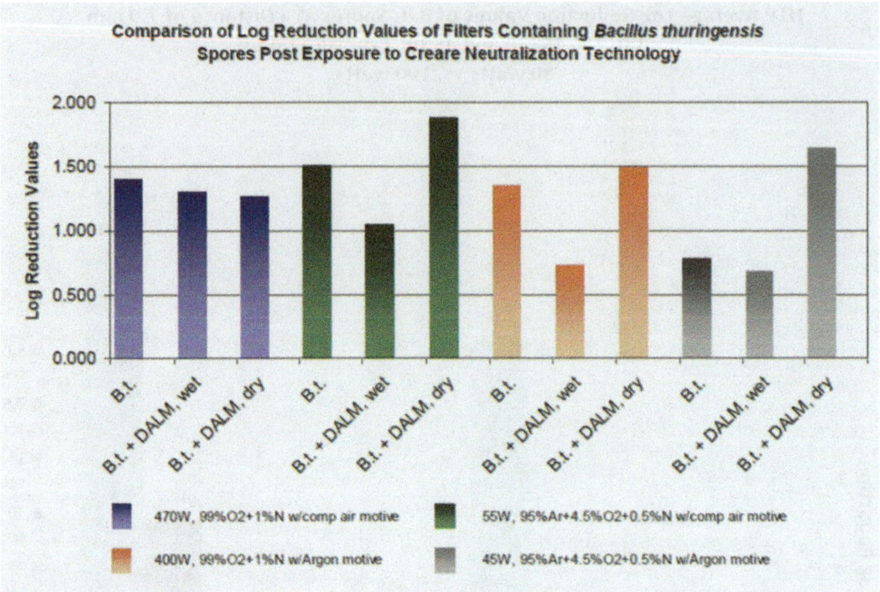

Fig. 8.22 The effects of DALM interactions with cold plasma intervention on cold plasma killing of Bt spores

Additional observations with applied plasma to bacterial spores coated with DALM nanoparticles support these conclusions. When a thin coat of bound water was present, it acted as a shield against cold plasma killing of DALM nanoparticle-coated spores, but when dry, the plasma-killing active gas species were able to kill the spores, albeit at a much lower efficiency. We now hypothesize that the water coating traps the reactive gases on the surface between the water and DALM metallic nanoparticles. This trapping allows them to be concentrated within this region even if they are very reactive, short-lived radicals such as hydroxyl radicals. The applied plasma gas was on the wrong side of the water coating until it was removed by drying. This interpretation also means that the activated bound water-coated DALM metallic nanoparticles must be in direct close contact with the target microbe to be effective in killing. The widespread bulk effect of killing by these nanoparticles is just a wide distribution of a very local effect. Microwave activation must be ongoing for the killing effect to continue, and any small distance from the nanoparticle source of plasma would rapidly extinguish the effect, with no collateral damage. The active gases generated and trapped in the nanobubbles on the surface of the particles are possibly hydroxyl radicals, monatomic oxygen, singlet oxygen, nitrogen oxide radicals, carbonate radicals, and/or formyl radicals. These remain to be experimentally determined.

One way to determine the active species would be to take the spectra of the emissions from visible through the near infrared to identify them in the air, with bound

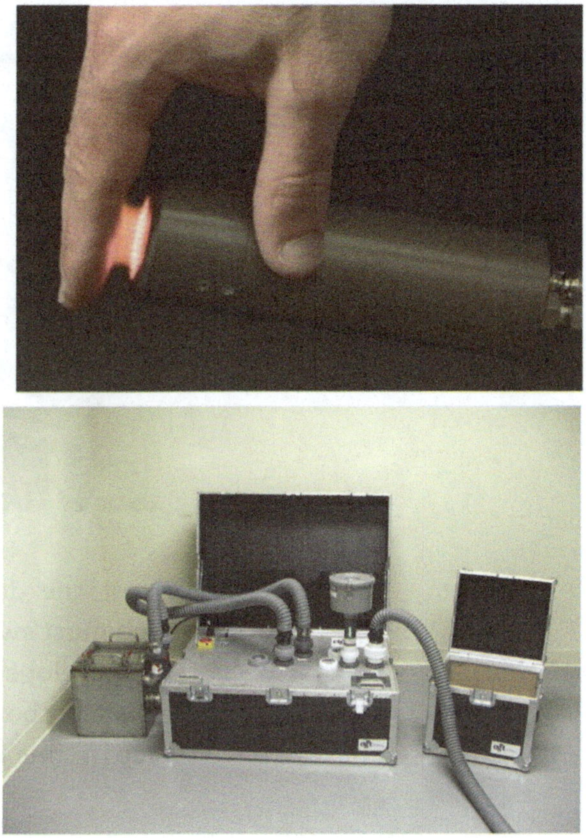

Fig. 8.23 Cold plasma devices (safe for decontaminating fragile surfaces, even human skin) for neutralizing biological agents on surfaces

water present, upon exposure to pulsed microwaves or RFR. The water absorption and self-absorption of the infrared should be minimized.

These DALM metallic nanoparticles can be dispersed by electronanosprayers (available in sizes from the size of a cell phone to a pen), which can distribute 1 gram of 20 nm particles over a volume of 10 m × 10 m × 1 km in 24 h at a concentration of over 2 million nanoparticles per cubic centimeter. This distribution makes the much-localized NEMR nanoparticle-mediated effect a wide-area large-scale application.

The assumption that NEMR was necessary for nanoparticles or carbon nanotubes to penetrate bacterial or even eukaryotic cells (animal and human) was soon disproven. This phenomenon occurred with DALM nanoparticles with or without added metallic cores and even without the addition of single-walled carbon nanotubes (SWCNTs). The penetration was not very clonal efficient, but large numbers of particles could overcome this limitation.

This "control effect" (for the lack of applied NEMR) was hard to explain. The application to human and animal cells will be discussed in the next chapter. Here, we will first approach the most disconcerting effect of genetically altering a

Fig. 8.24 Cold plasma devices (safe for decontaminating fragile surfaces, even human skin) for neutralizing biological agents on surfaces tested with a robotic applicator for precision application

To test the devices with applicators which must sweep a surface to kill, our team developed a test chamber with a Cartesian robot to precisely control the sweep in time and space:

Fig. 8.25 Cold plasma devices (safe for decontaminating fragile surfaces, even human skin) for neutralizing biological agents on surfaces showing close up of robotic applicator

The Atomflo™ 250 plasma applicator by Surfx Technologies above

pathogen, anthrax bacteria, with the very DALM nanoparticles meant to kill it. Additionally, of further concern was that it passed from species to species genetically and taxonomically remote from each other. From the first efforts to clone genes in expressible plasmids in *E. coli* to biosynthetically produce DALM, we observed that DALM had an affinity for bacterial DNA that made it copurify with the DNA but did not prevent the DNA's replication or transcription and translation. Two possibilities needed to be considered: (1) weak environmental NEMR facilitated the entry of the DALM nanoparticle DNA complexes, or (2) another mechanism, yet to be discovered, facilitated entry. Either could be problematic at best, especially not knowing what the latter could be. Unlike adding chemicals such as calcium phosphate or detergents or liposomes to DNA, the addition of divalent cations such as barium, calcium, or magnesium only made DALM into an insoluble precipitate. Lipophilic (water-miscible) solvents such as DMSO caused DALM to form colloidal globules (nanoparticles). However, the addition of metallic nanoparticles, such as those of iron, facilitated the entry of synthetic and biosynthetic nanoparticles into bacterial cells (Figs. 8.26 and 8.27).

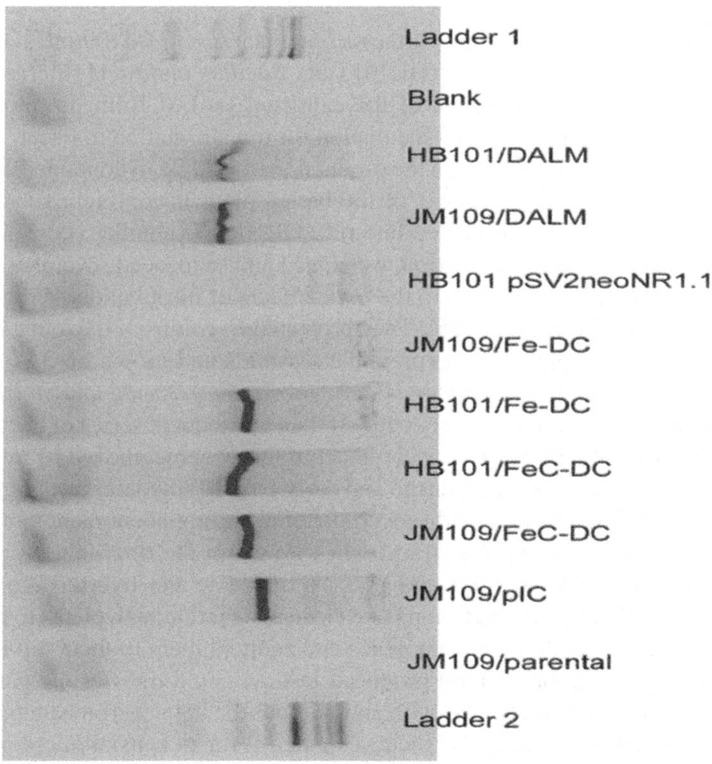

Fig. 8.26 This is the first example of genetic transfer between strains of *E. coli* by biosynthetic DALM nanoparticles

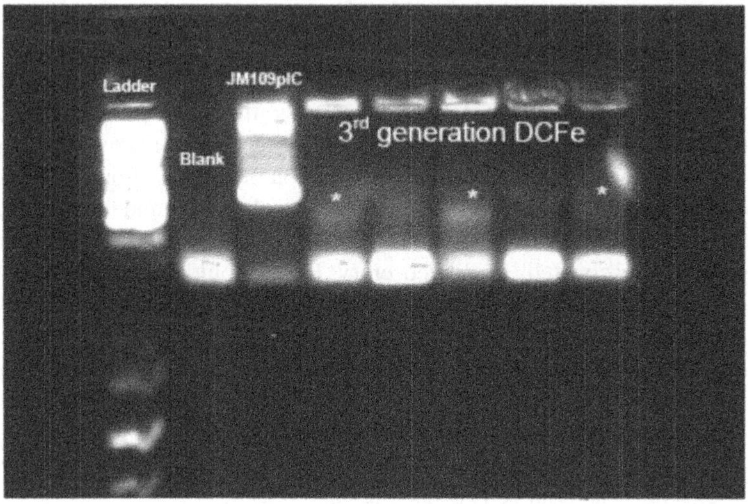

Fig. 8.27 PCR genetic evidence of transfer from transformed *E. coli* to another strain by the subsequent production of DALM nanoparticles for three iterations

For DALM nanoparticles made by the recombinant *E. coli* JM109/plC2ORNR1.1, entry into other strains of *E. coli* (HB101) and *Bacillus anthracis* (Sterne, vaccine strain) only required the addition of the extracted DALM from postbiosynthetic JM109 to these other bacteria and incubation for transformation to occur.

Although Ba was less prolific in the production of DALM, it contained sequences that hybridized with the sequences of the barley plant flavoprotein subunit of its nitrate reductase, which was cloned into plC2ORNR1.1 (number is the size of the cloned gene in kilobases). In addition, we cloned and sequenced several other genes from Ba, which made it resistant to the toxic effects of nitrite under aerobic conditions and DALM biosynthesis [15]. They protected *E. coli* from the lethal effects of 3AT medium when cloned into expressible plasmids and grown on 3AT medium. Other bacteria, such as *Pseudomonas aeruginosa*, *fluorescens*, and *stutzeri*, were also resistant to the lethal effects, but only if their subsequent series of enzymes that converted nitrite into nitrogen gas were intact. If these were removed by mutation or deletion, then the accumulating nitrite and early DALM oligomers would kill these species as well. The subsequent discovery that the coiling phenomenon of *Bacillus anthracis* was natural to all the strains when grown in the rhizospheres of certain plants [16], especially legumes and grasses, added to the hybridization results. These facts implied that Ba had a long and close association with plants to the extent that the bacteria had taken up plant genes and adapted them to their survival. PCR clearly demonstrated that JM109-produced DALM could transfer the genes of its origin from this strain to HB101, with and without the help of iron nanoparticles.

Other intracellular pathogens, such as spotted fever rickettsial microorganisms, could be collected by aptamers (specifically binding DNA to surface antigenic markers of the *Rickettsia*) that are connected to iron nanoparticles on microbeads

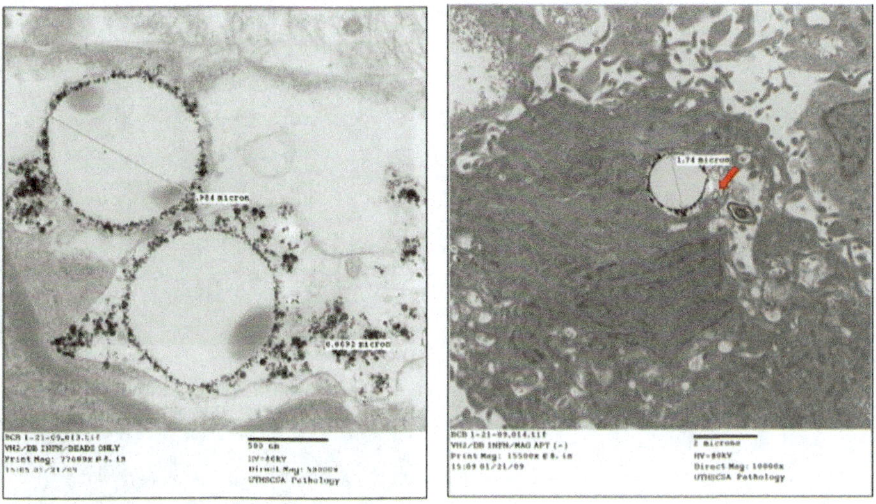

Fig. 8.28 Transfer of infectious live *Rickettsia* by metallic (iron) nanoparticles with antigen-specific aptamers attached to the particles and agents and, in turn, attached to microbeads; the arrow points to one with a single *Rickettsia* transferred to a vulnerable cell

and taken up by susceptible host cells, and could manifest a pure viable infection of such host cells (Figs. 8.28 and 8.29). Thus, such intracellular microbes can be tracked into their host cells and killed or made susceptible to other antimicrobials by DALM composite nanoparticles.

These data are an extension of the field collection and diagnostics of *Francisella tularensis,* tularemia bacteria, and the isolation and pure culture that we demonstrated in Houston, Texas (Figs. 8.31 and 8.32) [17].

Now, we have nanotechnology that can collect, identify, and lead to the pure culture of a pathogen from the contaminated field environment and that could prepare for seeking out and killing it in the environment.

The entry of these genes intact into the new host *E. coli*, mediated by DALM, was not the only or most remarkable result of these experiments. Each subsequent transformant could produce DALM/DNA nanoparticles that could transform another "generation" of *E. coli*, which, in turn, produced more "infecting nanoparticles" [18, 19]. This could occur in up to three subsequent cycles of production and transformation before playing out. When special aptamer sequences, made specifically against *Bacillus anthracis*, were imbedded in the pIC2ORNR1.1 plasmid and then produced along with DALM in the JM109 host, the biosynthetic nanoparticles produced could transform Sterne strain Ba (Fig. 8.33.). The new strain of recombinant Ba was not only ampicillin resistant (the marker and selection gene for successful clones) but also Cherry gamma phage resistant and hemolytic (Ba already contains silent hemolysis genes, but these were assumed activated because the

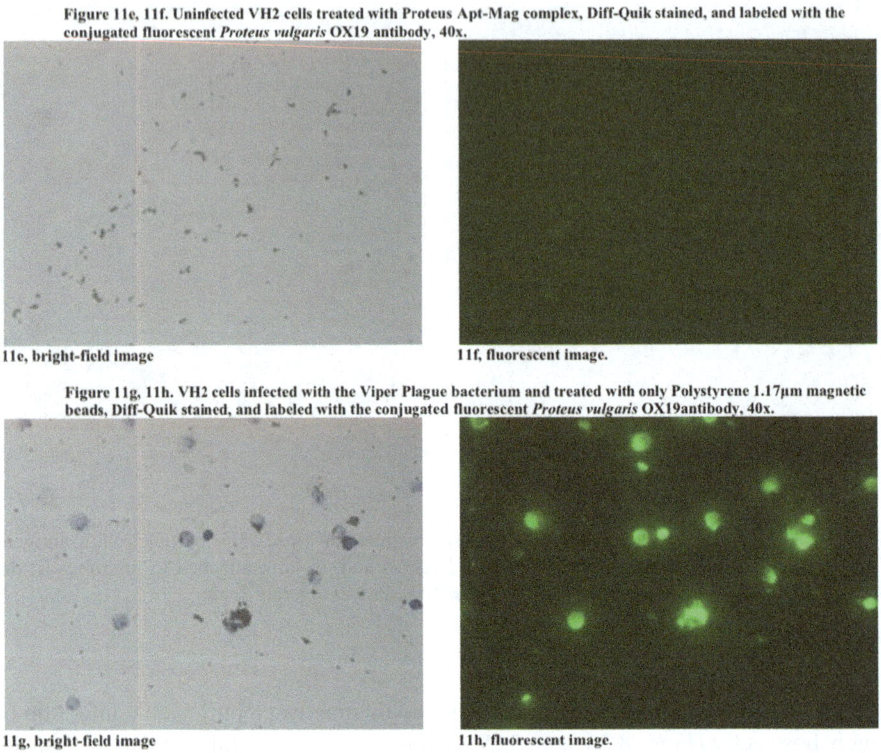

Fig. 8.29 Panels of visible light and fluorescence microscopic control images of fluorescent antibody-treated VH2 cells. The top panels are of uninfected VH2 cells (negative control); the middle, images of *Richettsia* infected VH2 cells that cross-reacts with the fluorescent antibody (positive); and the bottom images

introduced DALM nanoparticle and the source *E. coli* contained no such genes) (Fig. 8.34).

Table 8.2 reveals these properties and that they are shared by other closely related species and some "naturally" transformed field isolates, especially those recovered from Iraq after the first war and reportedly made by the Russians [20], in the open literature (although the latter was never reproduced or obtained from Russia).

Bacilli, in general, are known to obtain genes from temperate bacteriophages that infect but do not lyse the bacterial host and free DNA (albeit very inefficiently). However, this transformation by *E. coli*-derived DALM/plasmid DNA nanoparticles was very efficient in Ba. The PCR and electrophoretograms show that the specific genes were transferred to Sterne anthrax bacteria without contamination by other *E. coli* genes (Fig. 8.34).

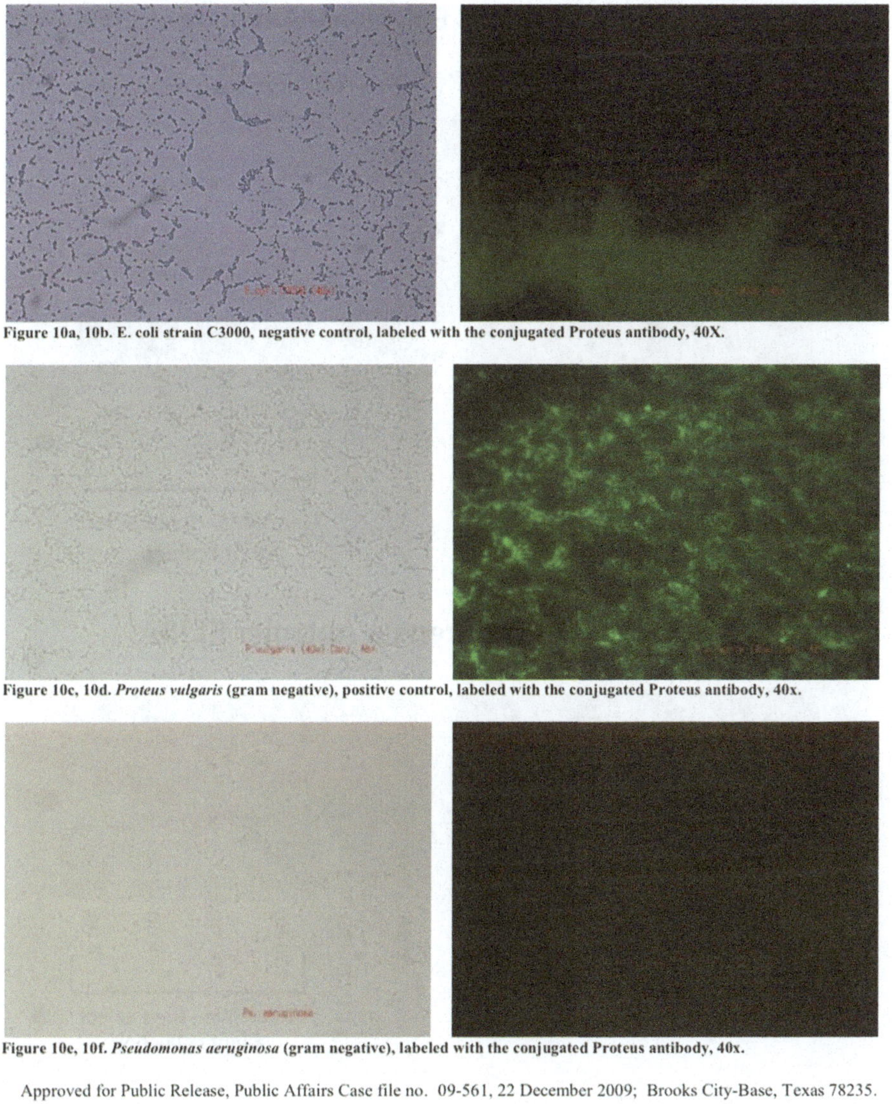

Fig. 8.30 Panels of visible light and fluorescence microscopic control images of fluorescent antibody-treated microbes. The top panels are of *E. coli* (negative control); the middle, images of *Proteus vulgaris* that cross-reacts with the fluorescent antibody (positive control); and the bottom images, *Pseudomonas aeruginosa* (negative control)

This was the next most important and essential step toward a synthetic to biosynthetic self-replicating nanoparticle (albeit with host cell being bacterial, at first, and then in the next chapter, eukaryotic host cell machinery help). Finally, the

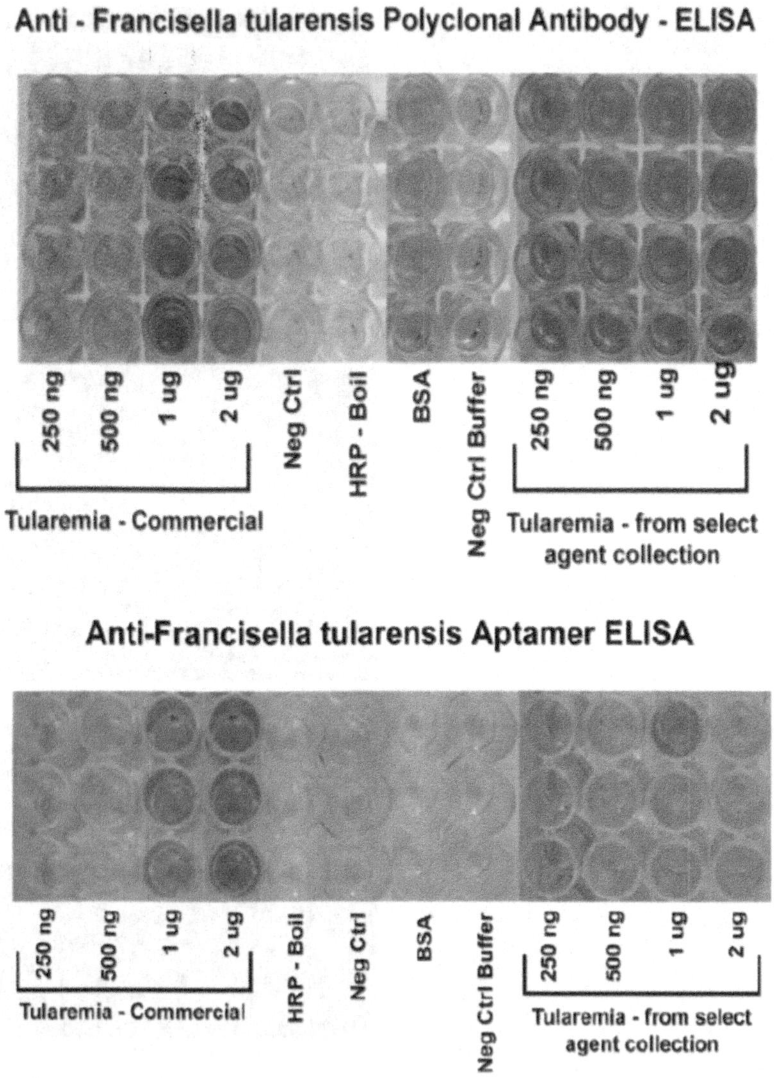

Fig. 8.31 Microtiter ELISA (enzyme-linked immunoassay) for *Francisella tularensis* (Ft), control antigens, and authentic Ft positive controls compared to an enzyme-linked aptamer microtiter plate test for Ft

DALM/DNA nanoparticles could be lyophilized and maintained at and in ambient temperatures and light, respectively, on the shelf in the laboratory for 4 plus years and remain functional (Fig. 8.35). Obviously, these results opened a wide variety of imagined and yet-to-be-imagined applications in molecular biology, even using the free radical reactions driven by DALM to specifically cut and deliver fresh new genes to modify the host cell's properties (as may have happened by accident in the Ba, where no new genes were added).

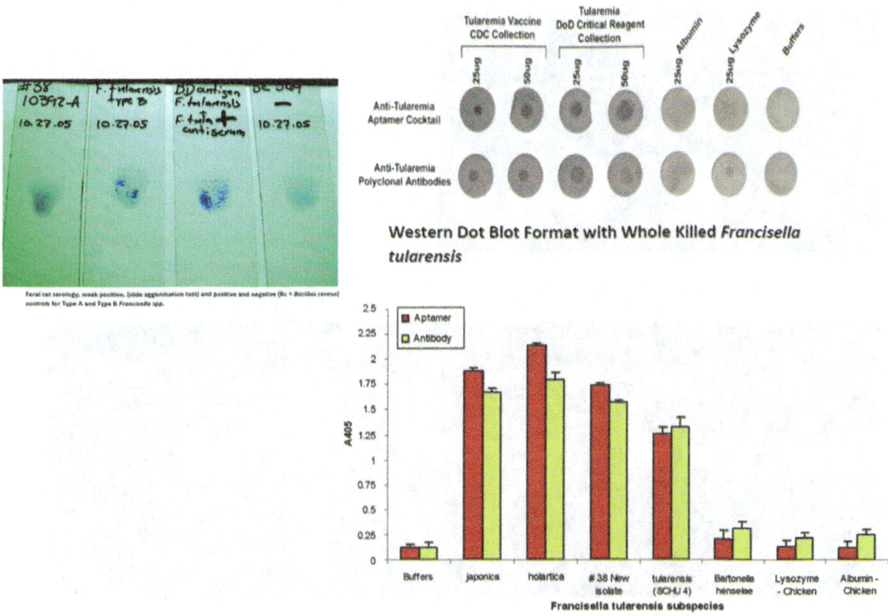

Fig. 8.32 Slide immunoagglutination test and comparison of antibody-based and aptamer-based dot blot results and quantitative ELISA results for identifying *Francisella tularensis* (tularemia)

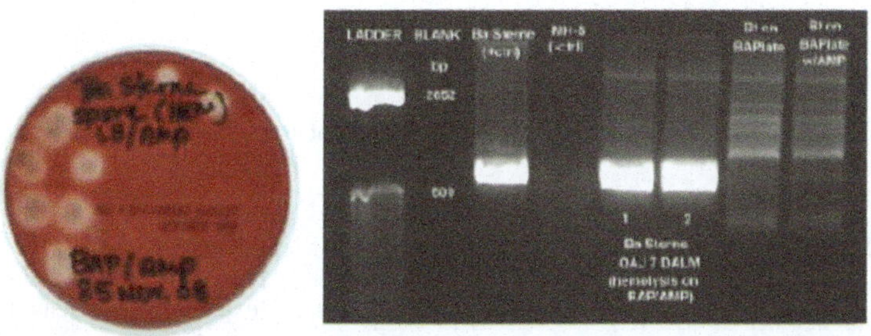

Species/strain	Ampicillin Sensitivity	Hemolysis	Cherry Gamma Phage Sensitivity	Temperate Phage (Expressed)	PlyG Sensitivity
Bacillus anthracis/OAJ7	-	+	-	-	+
Bacillus thuringiensis /Kurstaki	-	+	-	-	-

Fig. 8.33 *E. coli*-biosynthesized DALM nanoparticles transformed Sterne *Bacillus anthracis* to become ampicillin resistant, Cherry gamma bacteriophage resistant, and hemolytic, making it resemble a closely related nonpathogenic *Bacillus thuringiensis*

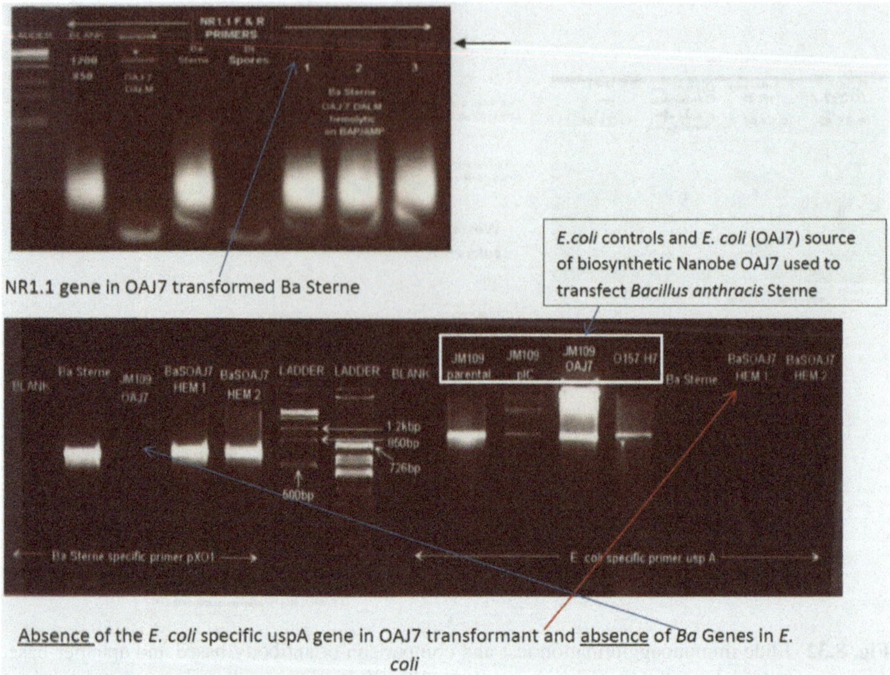

Fig. 8.34 PCR evidence that the transfer of traits from DALM-producing *E. coli* to *Bacillus anthracis* (Sterne) was not from bacterial cross-contamination

Table 8.2 Comparison of the characteristics of various strains of *Bacillus anthracis* and closely related *Bacillus* species

Species/strain	Ampicillin sensitivity	Hemolysis	Cherry gamma phage sensitivity	Temperate phage (expressed)	PlyG sensitivity
Bacillus anthracis/ OAJ7	–	+	–	–	+
Ba/K5A	+	–	–	+	+
Ba/K spore	+	–	+/–	–	+
Ba/KM10A	+	–	+/–	+	+
Ba/KM10B	+	+(temperature sensitivity)	+	+	+
Ba/KM12	+	–	+/–	–	+
Bacillus thuringiensis/ Kurstaki	–	+	–	–	–
Bacillus cereus/4342(ATCC)	+	++	+	+	+

Fig. 8.35 Highly stable lyophilized gene transferring DALM nanoparticles

References

1. Parker, J.E., Alls, J.L., and Kiel, J.L., U.S. Patent 5,902,728. Diazodenitrification in the Manufacture of Recombinant Bacterial Biosensors, 11 May 1999.
2. Kiel, J., Alls, J., Weber, R., and Parker, J., U.S. Patent 5,156,971. Rapid Identification of Environmental Bacillus, 20 Oct 92.
3. Kiel, J., Parker, J., Morales, P., Alls, J., Mason, P., Seaman, R., Mathur, S., and Holwitt, E. Pulsed microwave induced bioeffects. IEEE Transactions on Plasma Science 28(1): 161–167, 2000.
4. Norton, D. (Editor), Frog Embryo Teratogenesis Assay Xenopus (FETAX) for Soil Toxicity Screening, Pub No. 96-318, Department of Ecology, Washington State, April 1996.
5. Kiel, J., Parker, J. Holwitt, E., and Vivekananda, J., DNA capture elements for rapid detection and identification of biological agents. In Chemical and Biological Sensing V (P. J. Gardner, ed.), Proceedings of SPIE, vol. 5416, pp. 105–110, 2004.
6. Kiel, J., Parker, J., Mason, P., Holwitt, E., Stribling, L., Sloan, M., Morales, P., Vivekananda, J., Gonzalez, D., Tijerina, A., and Alls, J., Using Specific Binding DNA Capture Elements to Direct Pulsed Power Killing of Biological Agents, Digest of Technical Papers, 14th IEEE International Pulsed Power Conference (M. Giesselmann and A. Neuber, eds), vol. 1, IEEE, New Jersey, pp. 236–238, 2003.
7. Kiel, J., Holwitt, E., Parker, J., Vivekananda, J., and Franz, V., Nanoparticle-labeled DNA Capture Elements for Detection and Identification of Biological Agents. In Optically Based Biological and Chemical Sensing for Defence (J.C. Carrano and A. Zukauskas, eds.), Proceedings of SPIE, vol. 5617, pp. 382–387, 2004.
8. Kiel, J., Holwitt, E., Parker, J., Vivekananda, J., Franz, V., Sloan, M., Miziolek, A., DeLucia, Jr., F.C., Munson, C. A., and Mattley, Y.D., Specific biological agent taggants. In Chemical and Biological Sensing VI (P.J. Gardner, ed.), Proceedings of SPIE, vol. 5795, pp. 39–45, 2005.
9. Sloan, M., Vivekananda, J., Holwitt, E., and Kiel J., U.S. Patent 7,892,484. Methods and Compositions for Neutralizing Anthrax and Other Bioagents, 22 Feb 2011.
10. Sloan, M., Vivekananda, J., Holwitt, E., and Kiel, J., U.S. Patent No. 8,383,036. Methods for Neutralizing Anthrax or Anthrax Spores, February 26, 2013.
11. Jablonowski, H., et al., Quantification of the ozone and singlet delta oxygen produced in gas and liquid phases by a non-thermal atmospheric plasma with relevance for medical treatment, Scientific Reports 8:12195, https://doi.org/10.1038/s41598-018-30483-w, 2018.
12. Weldon, D., et al., Singlet Sigma: The "Other" Singlet Oxygen in Solution, Photochemistry and Photobiology 70(4): 369–379, 1999.

13. Gemmell, N., et al., Singlet oxygen luminescence detection with a fiber-coupled superconducting nanowire single-photon detector, OPTICS EXPRESS 21 (4): 5005–5013, 2013.
14. Aerssens, D., et al., A Photosensitized Singlet Oxygen ($^1O_2$) Toolbox for Bio-Organic Applications: Tailoring $^1O_2$ Generation for DNA and Protein Labelling, Targeting and Biosensing, Molecules 27: 778. https://doi.org/10.3390/molecules27030778, 2022.
15. Parker, J.E., and Kiel, J.L., US Patent 6,878,519 B1. DNA Probes for Specific Genes of Anthrax, 12 April 2005.
16. Saille, E., and Koehler, T.M., *Bacillus anthracis* Multiplication, Persistence, and Genetic Exchange in the Rhizosphere of Grass Plants, Applied and Environmental Microbiology, 72(5): 3168–3174, May 2006.
17. Vivekananda, J., and Kiel, J., Anti-Francisella tularensis DNA aptamers detect tularemia antigen from different subspecies by Aptamer-Linked Immobilized Sorbent Assay, Laboratory Investigation 86: 610–618, 2006.
18. Kiel, J. L., Parker, J.E., Holwitt, E.A., Vivekananda, J., Sloan, M.A., and Stribling, L.J.V. Progress in directed energy control of vectors for microbes and other cells. In Progress in Biomedical Optics and Imaging, Lasers in Surgery: Advanced Characterization, Therapeutics, and Systems XIV (Bartels, Bass, de Riese, Gregory, Hirschberg, Katzir, Kollias, Madsen, Malek, McNally-Heinzelman, Paulsen, Robinson, Tate, Trowers, Wong, eds.), Proceedings of SPIE, vol. 5312, pp. 326–333, 2004.
19. Kiel, J., et al. US Patent 8,628,955 B2. Compositions, Methods and Uses for Biosynthetic Plasmid Integrated Capture Elements, Jan 14, 2014.
20. Pomerantsev AP, Staritsin NA, Mockov YuV, Marinin LI. Expression of cereolysine AB genes in Bacillus anthracis vaccine strain ensures protection against experimental hemolytic anthrax infection. Vaccine. 1997 Dec;15(17–18):1846–50. https://doi.org/10.1016/s0264-410x(97)00132-1.

Chapter 9
Effects of the Biosynthesis of Radiofrequency and Microwave-Radiation-Absorbing Polymer on Eukaryotic Cells

Abstract The original reason for inserting a plant nitrate reductase gene in mammalian then human cells was to duplicate the diazoluminomelanin biosynthesis seen in bacteria. Initially, we thought certain cell lines could naturally produce diazoluminomelanin (DALM)-like luminescent polymers because they possessed inducible redox enzymes such as nicotinamide adenine diphosphate (NADPH) oxidase, nitric oxide synthase, tyrosine hydroxylase, and tyrosinase. HL-60 human granulocyte precursor (promyelocytes) cells and RAW 264.7 mouse monocytic/macrophage cells were the candidates, and other cell lines that were not expected to produce DALM were selected as controls. These two cell lines yielded luminescence. We went on to genetically introduce plant nitrate reductase into a variety of cell types. The gene not only enhanced or endowed the ability to produce DALM but also dramatically changed the phenotypes of these genetically modified cells. Finally, these changes and the additional genes introduced were tested for antiviral effects.

Keywords HL-60 · RAW 264.7 · NADPH oxidase · NO · Tyrosinase · Tyrosine hydroxylase · Vaccinia · Antiviral · Macrophage · Granulocyte · Cell lines · and ATCC

The first attempt to produce diazoluminomelanin (DALM) in eukaryotic cells was in mouse and human cells. B-16 mouse melanoma cells were a logical choice because they constitutively produce melanin by the action of tyrosine hydroxylase and tyrosinase, and DALM was considered, at that time, to be produced by the same pathways. The second choices had equal probability to produce DALM—HL-60 human promyelocytic cells and mouse monocyte/macrophage RAW 264.7 cells [1]. These cells can produce superoxide and peroxide by the action of inducible nicotinamide adenine diphosphate (NADPH) oxidase and/or the action of inducible nitric oxide synthase, which produces nitric oxide (NO) radicals and, subsequently, nitrite. They also contain myeloperoxidase, which was expected to catalyze redox reactions, forming DALM or melanin. HL-60 cells can be stimulated to produce superoxide and hydrogen peroxide via stimulation with phorbol-12-myristate-13-acetate (PMA), and RAW 264.7 cells can be stimulated to induce superoxide by means of

J. L. Kiel, *Bionanotechnology*, https://doi.org/10.1007/978-3-031-63006-4_9

lipopolysaccharides (LPSs), endotoxins, gamma interferons, and tumor necrosis factors (TNFs). Superoxide dismutase induced in these latter stimulated cells inhibits the subsequent stimulation of nitric oxide synthase (NOS) and NO production. Various substrates and stimulants were added to these cells, and pigmentation was noted and thermochemiluminescence measured. The various components and the results of thermochemiluminescence are noted in Table 9.1.

NOS produces NO from the cooxidation of L-arginine and NADPH (nicotinamide adenine diphosphate, reduced). The overall chemical equation is 2 L-arginine +3 NADPH +3H^+ + 4O_2, yielding 2 citrulline +2 NO + 4H_2O + 3NADP+. Therefore, NOS is a type of NADPH-dependent superoxide-producing oxidase, but the accumulation of hydrogen peroxide destroys the enzyme. Various combinations of 3-amino-L-tyrosine (3AT), nitrite, and luminol (L) were added to the cells to metabolize and produce DALM, as measured by hydrogen-peroxide-induced thermochemiluminescence. The production of pigment by HL-60 cells was expected from 3AT. The pigment was expected to be like melanin, that is, aminomelanin. DALM was assumed to be made from aminomelanin by the action of nitrite and luminol through a diazotization process. Unexpectedly, the mouse melanoma and macrophage cell lines produced less pigment than HL-60 cells. However, the macrophages produced the most luminescence under various experimental conditions. The highest luminescence for HL-60 cells was with nitrite and luminol in the absence of 3AT. The macrophages were also highest in luminescence under these conditions but still high and much higher under conditions for the production of complete DALM.

In addition to the production of DALM in these cell lines using existing metabolism, there were dramatic cellular effects that were harbingers of even more to come when the relevant metabolism was enhanced by genetic modification. HL-60 cells were greatly inhibited in growth in the presence of DALM precursors even though they remained viable.

These reported data led to the next step in DALM biosynthesis in eukaryotic cells, including human cells, the introduction of a plant nitrate reductase gene into those cells to enhance and control DALM biosynthesis [2, 3]. We were once asked by a reviewer why we chose to put a plant gene in animal and even human cells. The answer lies in the deep evolutionary relationship between nitrate reductase plant and bacterial genes and the nitrogen oxide generating genes of animals and humans. The results we are about to discuss bear out that confidence in this conclusion.

Table 9.1 In situ developed luminescence in established mammalian cell lines

Peroxide activated cellular luminescence (RLU)		
	3AT/NO_2/L	NO_2/L
Cell type		
B-16	464.4	423
HL-60	606.3	3395
RAW264.7	3109	4390

RLU relative light units, *3AT* 3-amino-L-tyrosine, *L* luminol, *NO_2* nitrite. (Based on data from reference [1])

The diagram of the plasmid vector we designed and used to transform a wide variety of cells to produce DALM is shown in Fig. 9.1.

First, we thought only monocyte/macrophage or granulocyte cells were suitable for stable and functional transfection and special episome promoters. We were soon to realize our thinking was too limited. We started with the RAW264.7 cells, P388D1 and J774A.1 mouse macrophages. They were transfected by electroporation. The graphic results are given in Figs. 9.2 and 9.3.

The P388D1 and RAW264.7 cells were successfully transfected, but the J774A.1 cells failed to demonstrate the functional uptake of the plasmid. The first change noted in successfully transfected cells was the slowing of replication compared to control cells without the loss of viability in addition to the production of nitrite in the presence of nitrate in the growth medium. The cells were delayed in the G2/M (growth phase 2 and mitotic phase interface) by the action of the modified nitrogen metabolism confirmed by cell cycle changes made by nitric oxide inhibitors and agonists [4].

The other change that showed that the introduced gene was incorporated into an existing redox enzyme was the production of additional nitrite with the addition of nitrate and the stimulation of nitric oxide synthase by LPS and gamma interferon (Fig. 9.4).

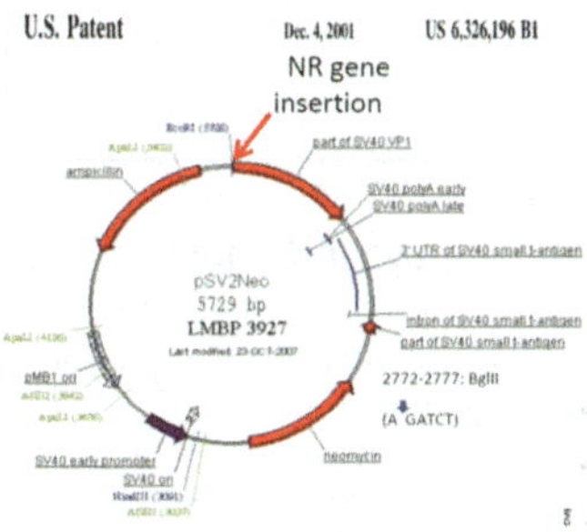

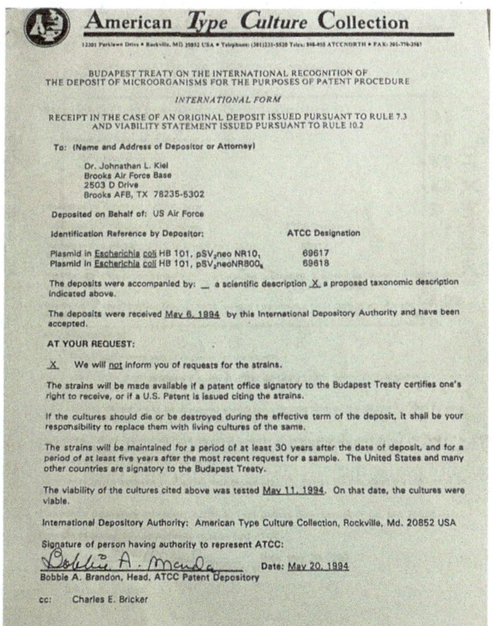

American Type Culture Collection

BUDAPEST TREATY ON THE INTERNATIONAL RECOGNITION OF THE DEPOSIT OF MICROORGANISMS FOR THE PURPOSES OF PATENT PROCEDURE

INTERNATIONAL FORM

RECEIPT IN THE CASE OF AN ORIGINAL DEPOSIT ISSUED PURSUANT TO RULE 7.3 AND VIABILITY STATEMENT ISSUED PURSUANT TO RULE 10.2

To: (Name and Address of Depositor or Attorney)

Dr. Johnathan L. Kiel
Brooks Air Force Base
2503 D Drive
Brooks AFB, TX 78235-5302

Deposited on Behalf of: US Air Force

Identification Reference by Depositor:	ATCC Designation
Plasmid in Escherichia coli HB 101, $pSV_2neo\ NR10_1$	69617
Plasmid in Escherichia coli HB 101, $pSV_2neoNR800_8$	69618

The deposits were accompanied by: __ a scientific description X a proposed taxonomic description indicated above.

The deposits were received May 6, 1994 by this International Depository Authority and have been accepted.

AT YOUR REQUEST:

X We will not inform you of requests for the strains.

The strains will be made available if a patent office signatory to the Budapest Treaty certifies one's right to receive, or if a U.S. Patent is issued citing the strains.

If the cultures should die or be destroyed during the effective term of the deposit, it shall be your responsibility to replace them with living cultures of the same.

The strains will be maintained for a period of at least 30 years after the date of deposit, and for a period of at least five years after the most recent request for a sample. The United States and many other countries are signatory to the Budapest Treaty.

The viability of the cultures cited above was tested May 11, 1994. On that date, the cultures were viable.

International Depository Authority: American Type Culture Collection, Rockville, Md. 20852 USA

Signature of person having authority to represent ATCC:

Date: May 20, 1994

Bobbie A. Brandon, Head, ATCC Patent Depository

cc: Charles E. Bricker

Fig. 9.1 Plasmid construct for transfecting human and animal cells with expressible nitrate reductase for DALM biosynthesis

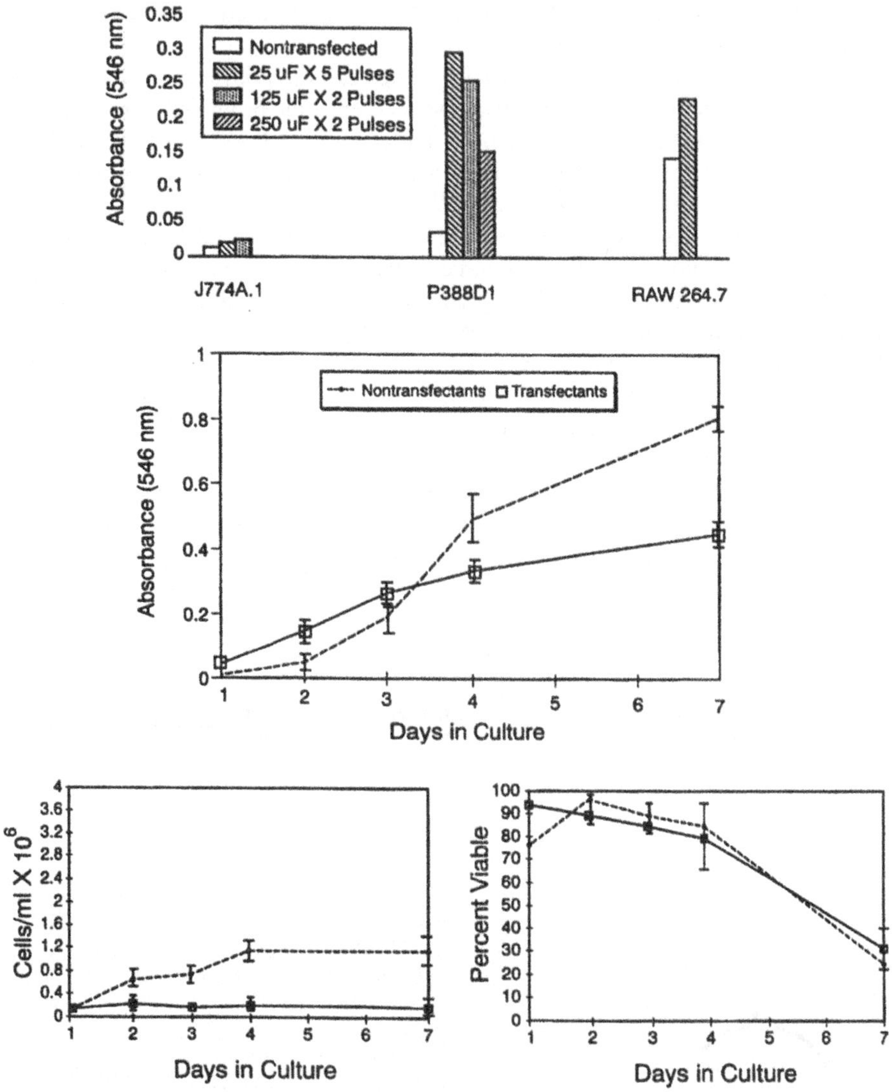

Fig. 9.2 Electroporation parameters and cell types for nitrate reductase expressible transfection. Additionally, the effects on growth and viability of successful transfection

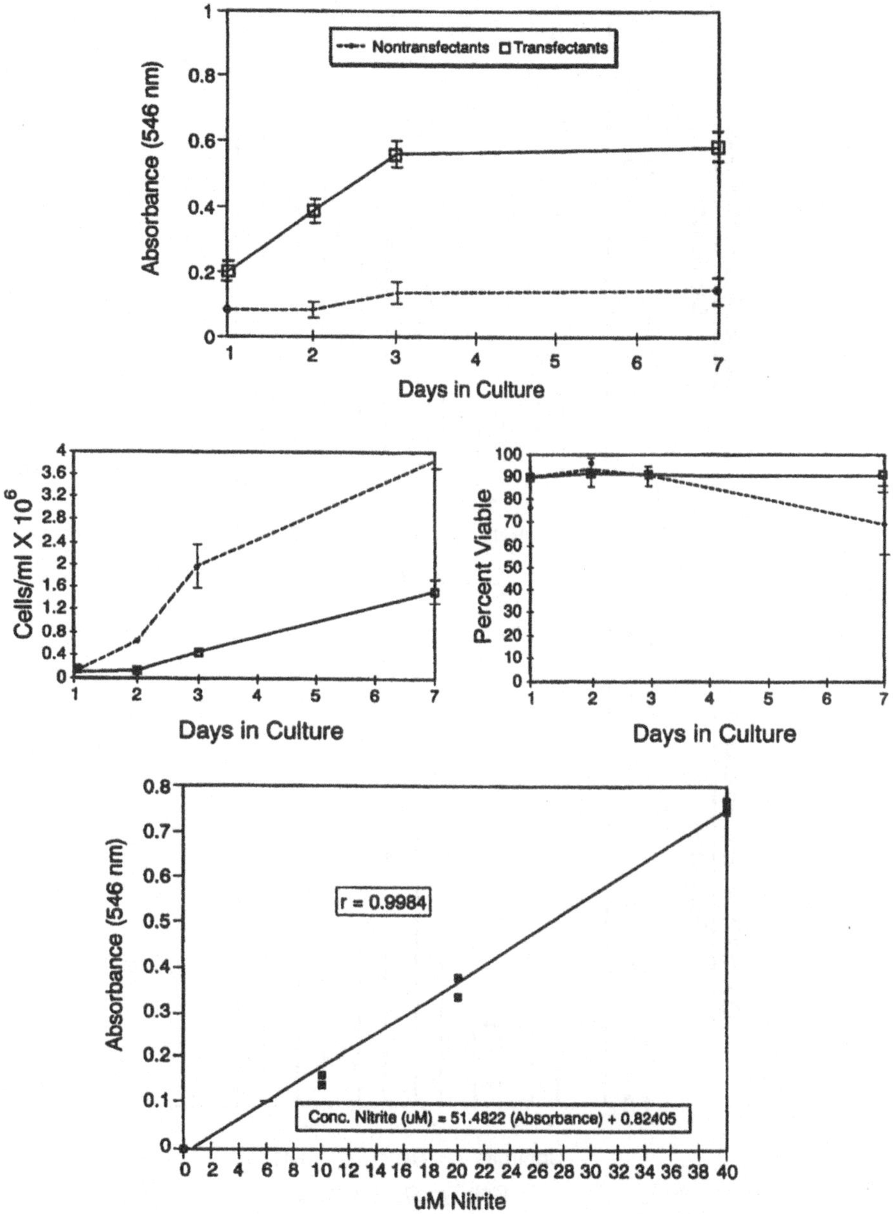

Fig. 9.3 Replication delays with sustained viability of plant-nitrate-reductase-transfected cells correlated with nitrite production

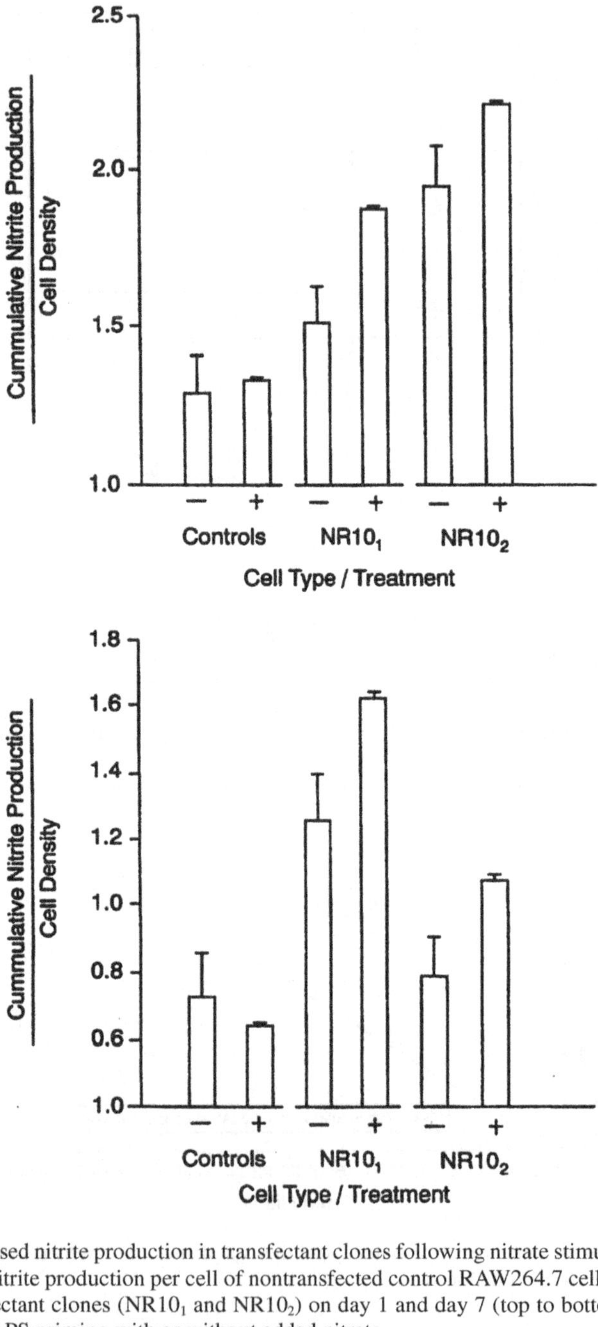

Fig. 9.4 Increased nitrite production in transfectant clones following nitrate stimulation compared to the relative nitrite production per cell of nontransfected control RAW264.7 cells and neomycin-resistant transfectant clones ($NR10_1$ and $NR10_2$) on day 1 and day 7 (top to bottom) post gamma interferon and LPS priming with or without added nitrate

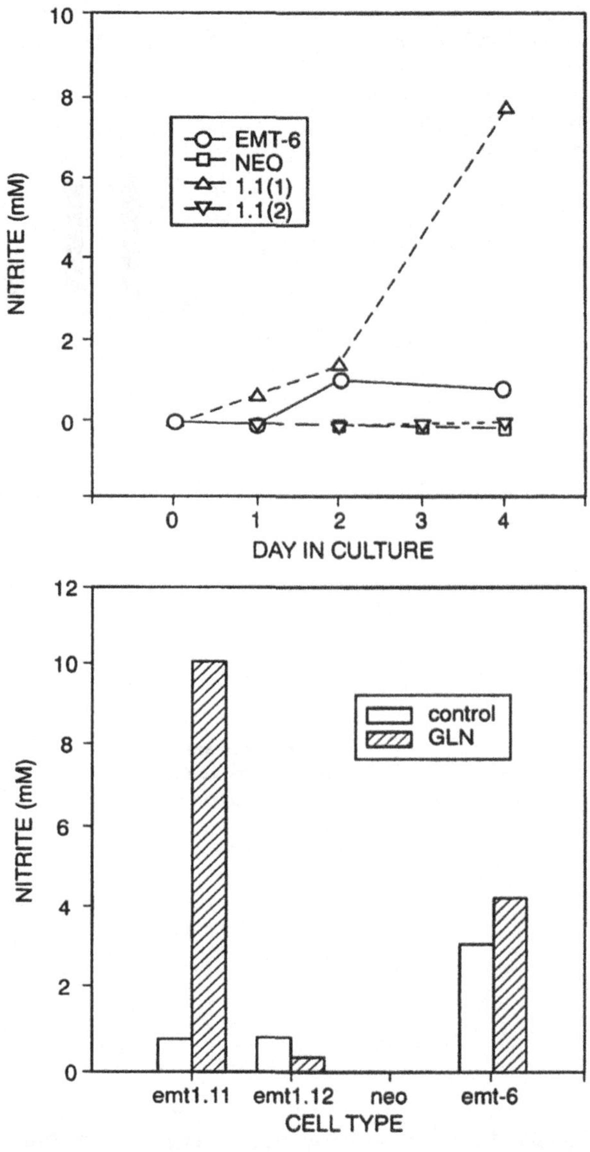

Fig. 9.5 Transfection of mouse mammary adenocarcinoma cells with plant nitrate reductase gene, showing enhanced nitrite production induced by gamma interferon, lipopolysaccharide, and nitrate. *GLN* gamma interferon, *L* lipopolysaccharide, and nitrate

The structural similarities of nitrate reductase and NOS appeared to allow them to share subunits and lead to additional combined, mixed functionalities. This mixed function was confirmed by NOS inhibitors blocking nitrite production from nitrate.

Similar results were obtained with the mouse adenocarcinoma mammary tumor cell EMT-6 (Figs. 9.5 and 9.6) [5].

Nitrite production was dependent on gamma interferon, LPS, and nitrate, with some small amount of constitutive breakthrough NO production. Two clones were produced that had different kinetics of nitrite production and a decrease in

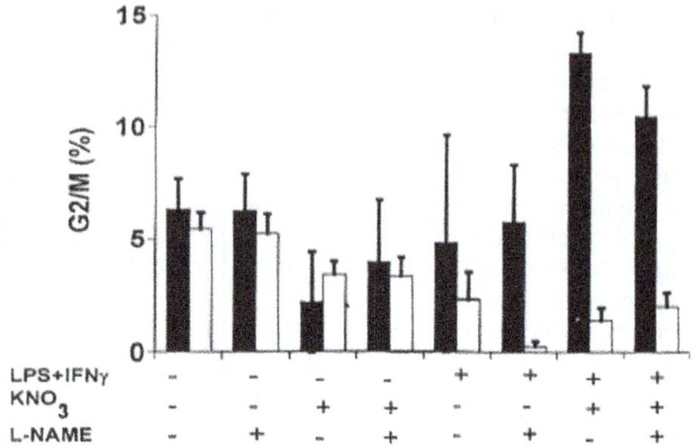

Fig. 9.6 Effects of nitrate reductase transfection and inhibitor and inducers of nitric oxide synthase and nitrate on cell cycle progression

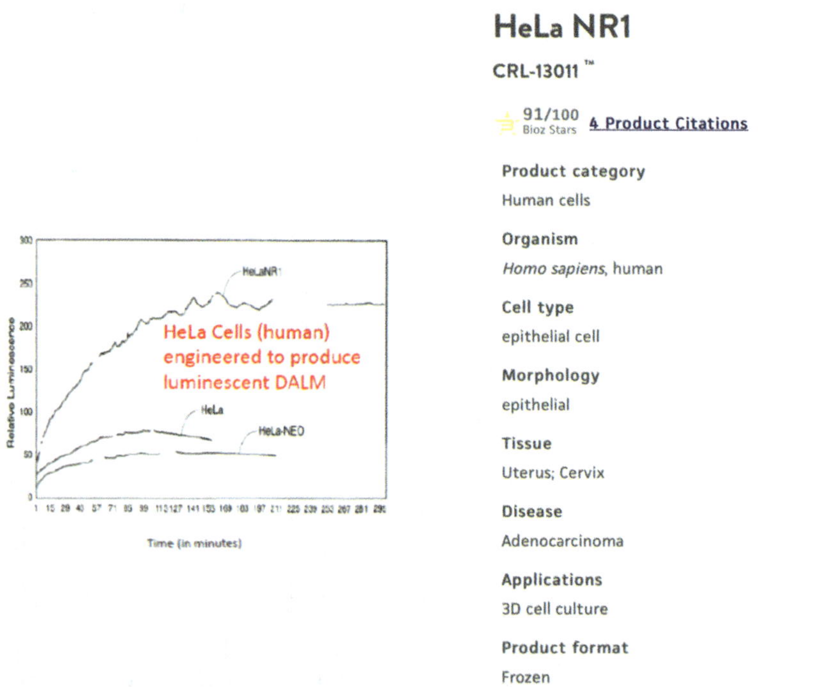

Fig. 9.7 Thermochemiluminescence of HeLa NR1 cells resulting from DALM biosynthesis

production over time. Other cell lines that were stably transfected to produce nitrite and DALM were HeLa (human cervical carcinoma cells induced by papillomavirus 18) and a lymphocyte cell line. HeLaNR1 CRL-12510™, ATCC, is a stable human cell line that produces DALM and natural nanoparticles (Figs. 9.7 and 9.10).

Other changes observed in transfected cell lines were morphological. The RAW264.7-transfected cells developed extremely long processes and greater mobility than the parent cell line. Some of these long processes connected cells (Figs. 9.8

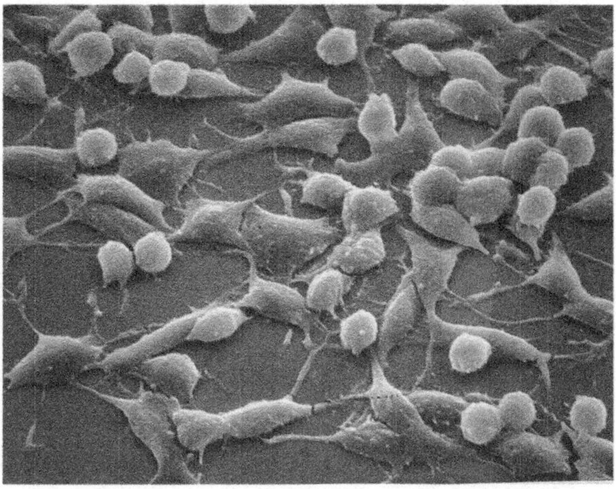

Fig. 9.8 Scanning electron microscopy of original RAW 264.7 cells

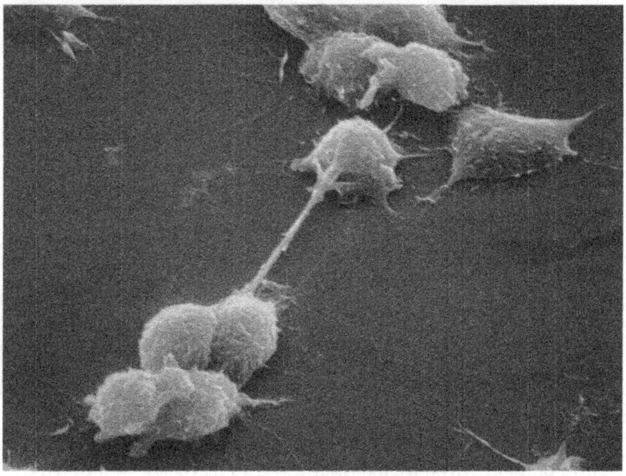

Fig. 9.9 Scanning electron microscopy of transfected RAW 264.7 cells showing exaggerated processes

and 9.9). In the parent cell line, when RAW cells were induced to convert from the round monocytic form to the attached macrophage form, they displayed shorter versions of these processes. When the RAW transfectants were treated with nitrite inducers, they collected in the rounder, loosely attached form, typical of cells accumulating in the G2/M cell cycle block (Fig. 9.8). HeLaNR1 cells, when fed DALM precursors, became more pleomorphic and accumulated electron-dense nanoparticles and nanoparticle clusters in the cytoplasm (Fig. 9.10) [6].

Eukaryotic cells in general were more resistant to the stasis and lethal effects of the introduced nitrate reductase gene that enhanced production of nitric oxide and nitrite even with the usual nitric oxide synthase inducers, TNF alpha and gamma interferon [7, 8].

The next logical step was to determine whether biosynthetic DALM-mediated gene transfer could occur across species among bacteria and eukaryotic cells and whether such genes could be expressed or cause the off-target activation of silent

Fig. 9.10 HeLa NR1 cell making DALM nanoparticles (white nanoparticles)

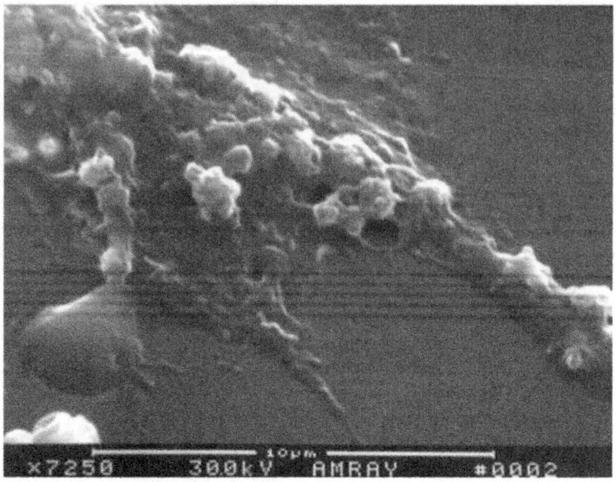

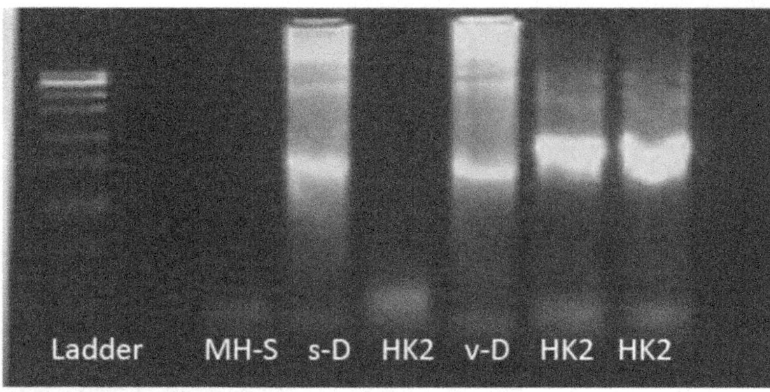

Fig. 9.11 PCR displaying DALM DNA transfer to human kidney cells (HK2) by biosynthetic and synthetic DALM/DNA nanoparticles. PCR of HK-2 and MH-S control cells is also shown

genes in the target host, as occurred in bacteria. Figure 9.11 is a polymerase chain reaction (PCR) electrophoretigram showing the NR plasmid uptake by human kidney cells (HK-2) mediated by both biosynthetic and synthetic DALM. The negative controls were the parent human kidney cell line and a parent mouse macrophage (MH-S) cell line.

PCR showed nitrate reductase genes (in plasmids) taken up with synthetic and biosynthetic DALM nanoparticles. This procedure was also successful with human lung epithelial cells (H441), as confirmed by PCR. Therefore, the utility of DALM nanoparticles for genetic engineering in a wide range of cells without using electroporation or ancillary chemical carriers was confirmed.

Aside from the production of nitrite and, subsequently, DALM, the real target was intracellular microbes such as obligate bacteria (*Rickettsia*) and viruses. The intracellular movement and processing of the DALM nanoparticles were traced by

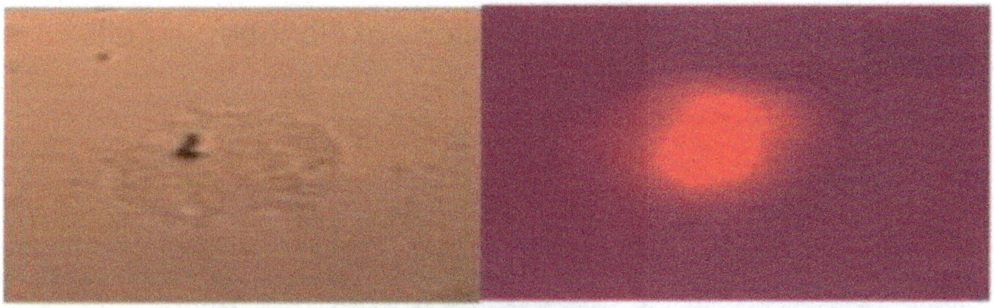

Fig. 9.12 HeLa-P6 cell line (human cells) showing the fluorescence of dequenched synthetic quenched double-stranded DNA aptamer and CNT nanoparticles (concentrating in the nucleus) after uptake

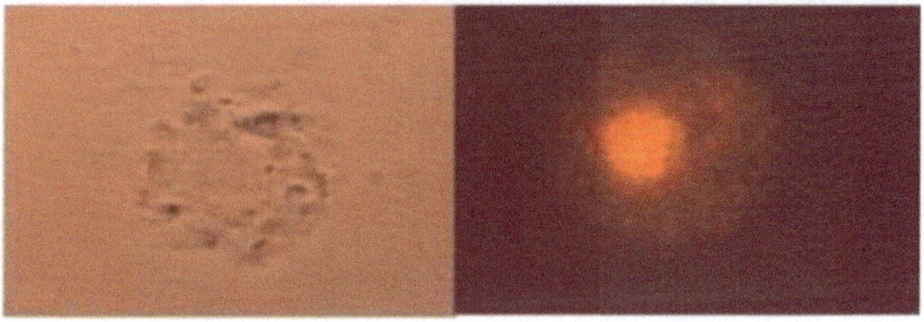

Fig. 9.13 Dequenched double-stranded DNA aptamer and CNT nanoparticles inside BS-C-1 cells (African green monkey cells) showing fluorescence

labeling with quantum dots or fluorescent dyes quenched by black hole quenchers or single-walled carbon nanotubes and metallic (iron) nanoparticles, which are electron dense for electron microscopy. These dequenched upon intracellular processing (Figs. 9.12, 9.13, and 9.14).

These techniques physically traced where the DALM particles went in the cells, such as in the nucleus or along cell membranes. The photomicrographs show the dequenching of the particles in the nucleus and carbon nanotubes, aligned by their long axes tangential to the membranes and the cytoplasmic clusters of iron nanotubes, all confirming the localization of certain components of these complex composite nanoparticles. However, this approach does not prevent unwanted off-target results. How could these be avoided and by what design?

The tracking of live pathogens into cellular hosts, as noted earlier for spotted fever rickettsia, has been demonstrated, but not the selective intracellular killing or inactivation of these pathogens. The first step to countering viruses inside host cells, using specifically directed antiviral technology that would be delivered by DALM nanoparticles, was investigated and successfully demonstrated just prior to the closing of the Brooks AFRL BSL3 laboratory in 2011.

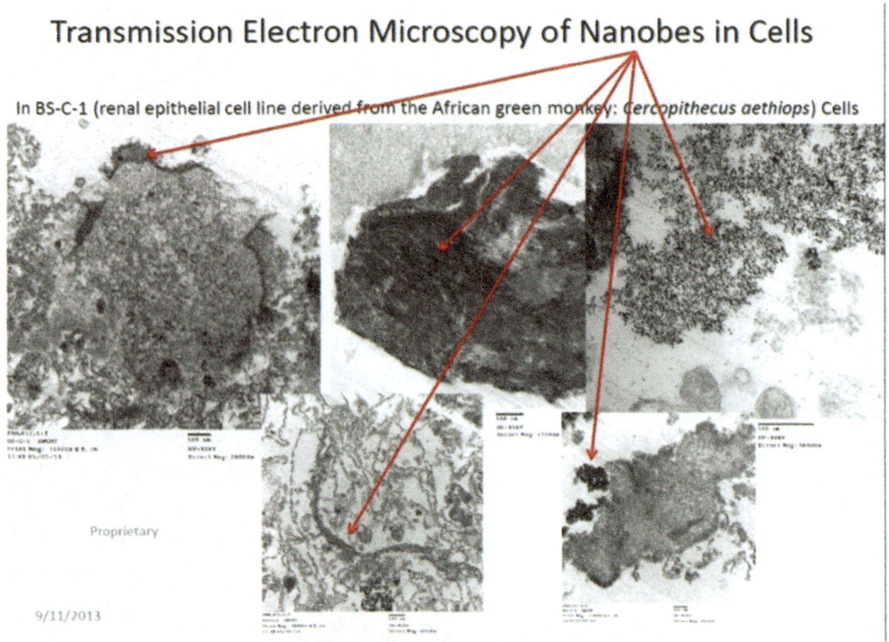

Fig. 9.14 Transmission electron microscopy of African green monkey cells displaying the uptake of carbon nanotube and iron nanoparticle combinations in membranes and cytoplasm

In RNAi (interference ribonucleic acid (RNA)) transcribed from deoxyribonucleic acid (DNA) cloned in plasmids, two vectors, pcDNA6.2-GW or pSV2neoNR1.1, inhibited the expression of the E3L gene in the vaccinia virus necessary for the replication of the virus in target HeLa cells (Fig. 9.15).

E3L gene proteins act as inhibitors of the double-stranded RNA (dsRNA)-dependent P1/eIF-2a kinase. A comparison of the assumed gene products of E3L and several dsRNA-binding proteins, including the dsRNA-dependent protein kinase, showed a shared motif involved in binding to dsRNA similar to that encoded by interferon-induced dsRNA-dependent protein kinase. The dsRNA-dependent kinase is an interferon-induced protein that phosphorylates the alpha subunit of the eukaryotic initiation elongation factor (elf)-2 complex, shutting off general translation, inducing apoptosis, and inhibiting viral replication.

Therefore, the effective plasmid removed the interferon-blocking effect of E3L and its antiapoptotic effect. When the cells were challenged with the vaccinia virus, the greatest statistically significant reduction in vaccinia viral plaques was not with the intact RNAi produced by plasmid DNA but by the nitrate reductase gene expressed with scrambled RNAi. Vaccinia virus was used as a surrogate for the pathogen smallpox and, now, mpox. In five plasmid combinations, the plasmids, whether having functional RNAi or not, favored cell lysis by the virus. This response did not mean that the virus successfully reproduced in these cells, only that viral-mediated cell lysis was enhanced, which could have been the result, at least in part,

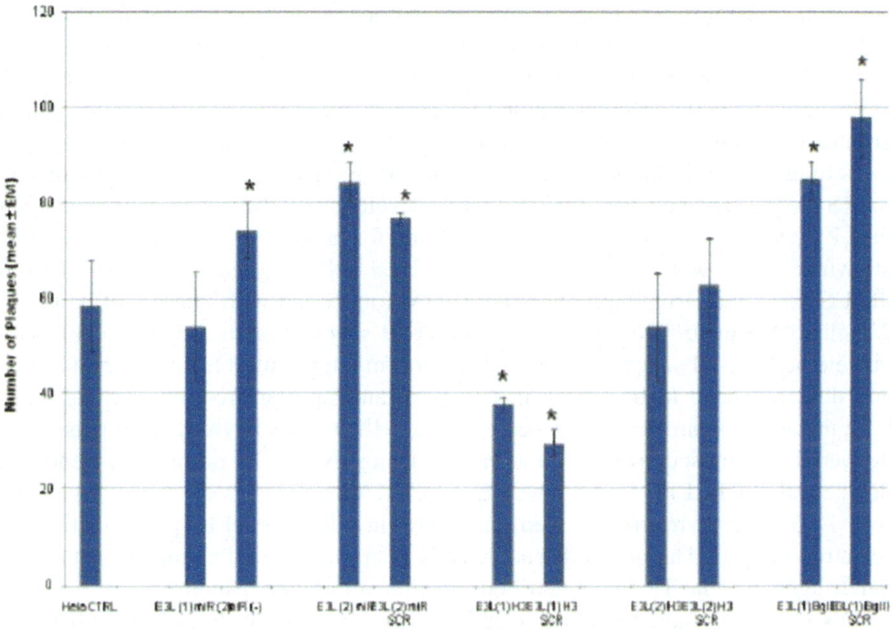

Fig. 9.15 Effects of the nitrate reductase carrying plasmids with and without interfering RNAi, which was intended to block vaccinia virus (a stand-in for smallpox virus). The most significant inhibition of viral replication was with the nitrate reductase gene alone. 1.0 MOI (multiplicity of infection) of vaccinia virus was used. The asterisk indicates significant differences between the HeLa control and the experimental samples. RNAi was cloned into the pcDNA6.2-GW vector (E3L(2)miR), and the paired scrambled control (miR SCR) increased the number of plaques compared to the HeLa control. These data suggest that the RNA exhibited E3L-independent effects in HeLa cells, which resulted in greater vaccinia virus pathogenicity. The E3L(1) and paired scrambled control were cloned into the DALM-producing plasmid pSV2neoNR1.1 vector using two different restriction enzyme sites, Hindill (H3) and BglII

of stimulating the apoptosis of the cells, even without complete viral replication. Oddly enough, when the DALM-producing pSV2neoNR1.1 plasmid (it also produced nitrite and nitric oxide) was used, whether with scrambled RNAi or not, viral plaque formation was significantly inhibited. This meant that NR1.1 probably had a nonspecific viral inhibiting effect, perhaps by delaying cell cycle progression, similar to what we had observed before with the nitrate reductase gene enhancement of nitrite production, preventing viral replication [9–12].

Recently, viral pathology (ferroptosis) has been linked to free radical damage [13], although the regulation of antioxidant enzymes by nitric oxide can be inhibitory or, in some cases, promoting or neutral [14]. The localization of the iron ion to certain motifs of DNA may have mediated double-strand breaking at purine-T-G-purine (RTGR). The subsequent repairing of these breaks may be relevant to chromosomal insertion of nitrate reductase genes and modification of the hemolysis gene and bacteriophage susceptibility of *Bacillus anthracis* by DALM [15]. It is a powerful chelator of iron ions and other transition metals. Iron is a ubiquitous metal in biology regardless of whether it is added to preparations or not.

The DALM nanoparticle technology could have had great utility in delivering DNA-based COVID vaccines beyond liposomes and in vivo electroporation. India granted emergency use authorization to a DNA COVID-19 vaccine that uses plasmids to prime the immune system by synthesizing protein antigens in situ against SARS-CoV-2. The vaccine, ZyCoV-D, was developed by the Indian pharmaceutical firm Zydus Cadila; it is the first DNA vaccine for humans to receive authorization anywhere in the world. On November 16, 2020, INOVIO received approval from the US Food and Drug Administration (FDA) to proceed with Phase 2 of its Phase 2/3 clinical trial for INO-4800, its COVID-19 vaccine candidate. INO-4800 is a nucleic-acid-based vaccine that is stable at room temperature for more than a year and does not need to be frozen in transport, an important requirement for mass immunizations in areas with limited resources. INOVIO's DNA vaccine plasmid is delivered intramuscularly or intradermally using INOVIO's proprietary handheld device called CELLECTRA®. The CELLECTRA device uses electrical direct current (DC) pulses to reversibly open small pores in cells in order for plasmids to enter (electroporation). The plasmids enable cells to produce specific antigens. The antigen triggers normal T-cell and antibody-mediated immune responses.

DNA vaccines have been criticized in the past because of the possible off-target effects on protein expression, the integration of foreign DNA into the host genome, and difficulties in delivery to the right target cells. INOVIO has provided evidence that it has overcome these limitations through plasmid design and a proprietary electroporation device. The concern about plasmids in comparison to messenger RNA (mRNA) vaccines is that they must enter the nucleus to be converted into mRNA before being translated into protein in the ribosomes of the cytoplasm. mRNA vaccines do not require entry into the nucleus before this last step. I have spent a lot of time providing accounts and data of the Brooks Counterproliferation Team's work on these problems for different military applications in the past. It is ironic that INOVIO has received a significant level of funding from the Department of Defense independent of our work: on June 23, 2020, INOVIO received a $71 million contract from the US Department of Defense to scale up its manufacture of its CELLECTRA® 3PSP Smart Device and procurement of CELLECTRA® 2000 for COVID-19 DNA vaccine delivery.

INOVIO's DNA vaccine clinical trials were terminated for undisclosed reasons. The CELLECTRA® electroporation device developed by Inovio Pharmaceuticals, Inc., is being assessed for prophylactic and therapeutic applications. For DALM nanoparticles, the use in the delivery of plasmid vaccines was an opportunity lost but perhaps only delayed.

The data are presented here before moving on to the subject of fully synthetic self-replicating antimicrobial nanoparticles because they represent historical evidence of progress toward that end and the collateral effects that must be understood and addressed to achieve it.

References

1. Bruno, J., and Kiel, J., Synthesis of Diazoluminomelanin (DALM) in HL-60 Cells for Possible Use as a Cellular-Level Microwave Dosimeter. Bioelectromagnetics 15: 315–328, 1994.
2. Kiel, J., Parker, J., and Bruno, J., U.S. Patent 5,464,768 (AF Inv 20,635) Enhanced Nitrite Production in Transfected Murine Cells, 7 Nov 95.
3. Bruno, J., Parker, J., and Kiel, J., Plant Nitrate Reductase Gene Fragments Enhance Nitrite Production in Activated Murine Macrophage Cell Lines. Biochemical and Biophysical Research Communications 201: 284–289, 1994.
4. Kalns, J., Parker, J., Bruno, J., Holwitt, E., Piepmeier, E., and Kiel, J., Nitrate Reductase Alters 3-Nitrotyrosine Accumulation and Cell Cycle Progression in LPS + IFN-γ-Stimulated RAW 264.7 Cells. NITRIC OXIDE: Biology and Chemistry 2(5): 366-374, 1998.
5. Parker, J.., and Kiel, J. U.S. Patent 6,013,520, Breast Tumor Cells for Nonionizing Radiation Effects, 11 Jan 2000.
6. Parker, J.E, and Kiel, J.L. U.S. Patent 6,326,196, Nitrate Reductase-transfected HeLa Cells for Cancer and Microwave Bioeffects Research, 4 Dec 2001.
7. Tuteja, N., et al., Nitric Oxide as a Unique Bioactive Signaling Messenger in Physiology and Pathophysiology, Journal of Biomedicine and Biotechnology 2004:4, 227–237
8. Taha Salim, T., Sershen, C., and May, E., Investigating the Role of TNF-α and IFN-γ Activation on the Dynamics of iNOS Gene Expression in LPS Stimulated Macrophages, PLOS ONE, https://doi.org/10.1371/journal.pone.0153289, June 8, 2016.
9. Lee, J., et al., Differential regulation of inducible nitric oxide synthase and cyclooxygenase-2 expression by superoxide dismutase in lipopolysaccharide stimulated RAW 264.7 cells, EXPERIMENTAL and MOLECULAR MEDICINE 41(9): 629–637, September 2009.
10. Ahmadabad, H., et al, Cytokine profile and nitric oxide levels in peritoneal macrophages of BALB/c mice exposed to the fucose-mannose ligand of Leishmania infantum combined with glycyrrhizin, Parasites Vectors 13:363, https://doi.org/10.1186/s13071-020-04243-7, 2020.
11. Radi, R., Peroxynitrite, a Stealthy Biological Oxidant, JOURNAL OF BIOLOGICAL CHEMISTRY 288(37): 26464–26472, SEPTEMBER 13, 2013.
12. van der Vliet, A., et al., Aromatic hydroxylation and nitration of phenylalanine and tyrosine by peroxynitrite: Evidence for hydroxyl radical production from peroxynitrite, FEBS Letters 339: 89–92, 1994.
13. Chen, J.et al. Free Radical and Viral Infection: A Review from the Perspective of Ferroptosis. Vet.Sci.2023,10,456. https://doi.org/10.3390/vetsci10070456, 11 July 2023.
14. Begara-Morales JC, et al. Antioxidant Systems are Regulated by Nitric Oxide-Mediated Post-translational Modifications (NO-PTMs). Front. Plant Sci. 7:152. doi: https://doi.org/10.3389/fpls.2016.00152, 2016.
15. Rai, P., Cole, T.D., Wemmer, D.E., and Linn, S., Localization of Fe(2+) at an RTGR sequence within a DNA duplex explains preferential cleavage by Fe(2+) and H2O2. J. Mol. Biol. 312(5): 1089–101, 2001 Oct 5. https://doi.org/10.1006/jmbi.2001.5010. PMD: 11580252.

Chapter 10
Moving Toward a Totally Synthetic Autonomous Nanoparticle

Abstract The driver for developing a totally synthetic self-replicating autonomous nanoparticle that duplicated the functions of the genetically modified biosynthetic ones was not totally scientific. The military requirement grew out of the necessity to avoid the prohibition under international law, the treaty against the development of biological weapons, and the domestic bioterrorism act. Even though such nanoparticles were a countermeasure against bioweapons, they could be turned into them. The other pragmatic reason that stood in the way of the application of the technology was the collateral, off-target, and potentially harmful effects. This chapter addresses the steps that have been taken for the conversion to a fully synthetic form that replicates a biosynthetic one. The collateral effects have not been fully investigated, but at least some of them may be worth retaining for the potentially useful positive biomedical effects discussed.

Keywords Plasmid · Virus · Carbon nanotubes · Biocompatible · Nanoparticles · Autonomous · Aptamers

The drive to replace the biosynthetic diazoluminomelanin (DALM) nanoparticles with wholly synthetic nanoparticles that perform the same biodefense counterproliferation missions was to avoid a rising concern in the Department of Defense that biotechnology countermeasures might be misconstrued as dual-use weapons. Chinese military science pursuits in radiofrequency radiation (RFR)-controlled biotechnology reenforced these concerns. Therefore, even the perception of such dual use as an offensive weapon needs to be avoided.

The other drive to pursue this synthetic form was to minimize off-target effects by simplifying the DALM nanoparticle constructs and making them more chemically and physically defined. The foundation for accomplishing this was tied to the potential biochemical and physiological effects of nitric oxide (NO) and nitrite [1, 2]. NO can modify proteins by nitrosating their thiol groups to alter their function. For instance, the S-nitrosation of kinases usually inhibits their activity. The S-nitrosation of a thiol in c-Jun N-terminal kinase (JNK) inhibits its phosphorylation and activation of JUN, and nitrosation inhibits IκB kinase β (IKKβ), which

J. L. Kiel, *Bionanotechnology*, https://doi.org/10.1007/978-3-031-63006-4_10

activates IKK following tumor necrosis factor-alpha (TNF-alpha) stimulation. Other enzyme activities inhibited by the nitrosation of thiols are those of the tyrosine kinase activity of epidermal growth factor receptor (EGFR) and protein kinase C. The S-nitrosation of a cysteinyl residue of apoptosis signal-regulating kinase 1 (ASK1) inhibits its binding to its substrates, mitogen-activated protein kinase kinase 3 (MKK3) and MKK6, in turn inhibiting apoptosis. ASK1 is of the mitogen-activated protein kinase (MAPK) kinase kinase family and activates the MAPK kinase 4 (MKK4)/MKK7-JNK and MKK3/6-p38 pathways. These pathways are activated by oxidative stress, TNF alpha, lipopolysaccharides (LPSs), extracellular adenosine 5′-triphosphate (ATP), endoplasmic reticulum stress of forming misfolded proteins, and high levels of intracellular calcium.

Nitrite (and therefore nitric oxide and peroxynitrite) can cause mutations, which rise with increasing nitrite concentration and decreasing pH. The most commonly induced mutations were GC to AT, followed by GC to TA transversions. Additionally, peroxynitrite forms 8-nitroguanine, which leads to its spontaneous depurination from deoxyribonucleic acid (DNA) (Fig. 10.1). This reaction leads to single-strand breaks in DNA mediated by the repair enzyme DNA (apurinic site) lyase [3]. These biochemical reactions may explain some of the effects of introducing nitrate reductase genes into cells in culture.

NO can also bind to heme as in guanosine-3′,5′-monophosphate (GMP) cyclase, leading to its production of cyclic GMP (cGMP), which lowers intracellular calcium ions, causing smooth muscle relaxation and subsequent vasodilation. Therefore, NO regulates blood pressure, prevents the aggregation and adhesion of platelets, mediates the immune system killing of many pathogens, inhibits viral replication, promotes certain types of cancer, promotes penile erection and spermatogenesis, and facilitates pregnancy by increasing uterine blood circulation and quiescence by smooth muscle relaxation. Introducing nanoparticles into an entire host could therefore have profound pathophysiological effects.

Therefore, how to deliver to specific targets in a host and how to control the activity and replication of a synthetic DALM nanoparticle became paramount if it were to have any practical, safe use. Unlike technologies such as clustered regularly interspaced short palindromic repeats (CRISPR), which can cause off-target genetic effects without a fail-safe or way to call it back, the designs of the DALM

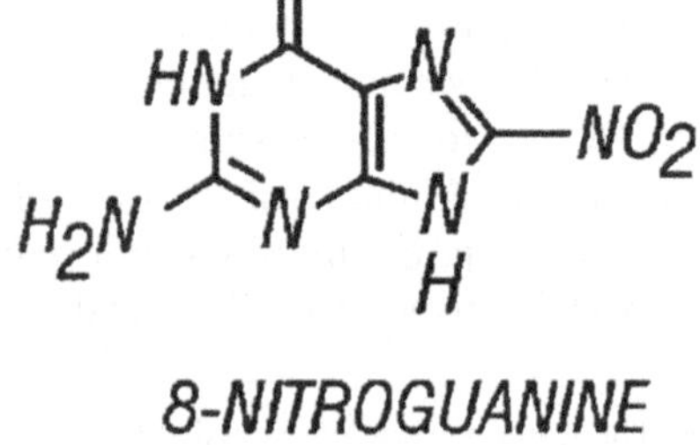

Fig. 10.1 Structure of 8-nitroguanine, the depurination product of the respective nucleotide from treating DNA with peroxynitrite [3]

nanoparticles required a fail-safe destruction mechanism, especially since it could lead to replicating and "infecting" progeny.

For comparison, attempts to use self-replicating vaccines have produced some untoward effects. A recent example is found in a very important agricultural disease of swine spreading in a worldwide pandemic: African swine fever (ASF). A recent study conducted by Iowa State University showed that an ASF outbreak in the United States would immediately close international markets to US pork and lower US live hog prices by 40%–50%. The US pork industry would lose between 15 billion and 50 billion dollars and 140,000 jobs within 10 years.

Recently, 1000 pigs died in Vietnam in the summer of 2022 after receiving the first commercial vaccine to prevent African swine fever (ASF). The pigs that received the NAVET-ASFVAC vaccine, codeveloped by the Vietnamese Navetco company and the US Agricultural Research Institute (ARS), part of the US Department of Agriculture (USDA), were on a farm in the central province of Phu Yen and two other provinces. Some of the pigs on a farm in Hoa Thinh Dong Commune in Phu Hoa District died a few days after receiving the vaccine. These were all healthy pigs that developed fever and hemorrhages, leading to many deaths.

The vaccine is of the next generation of self-amplifying messenger ribonucleic acid (saRNA) vaccines. They "copy" themselves after entering the animal and are anticipated to be of lower cost because smaller doses are needed for a protective response, and storage is easier (ProMed, PRO/AH/EDR, African swine fever—Asia Viet Nam, spread, vaccination trials, RFI, a program of the International Society for Infectious Diseases http://www.isid.org, October 15, 2022).

Self-amplifying RNA vaccines contain a transgene or "gene of interest" inserted into the vaccine platform, inducing an antibody response without the need for whole pathogens. Genvax is the US producer of the vaccine. They claim that self-amplifying RNA shows enhanced antigen expression at lower doses compared to conventional mRNA, suggesting that this technology may improve immunization. However, the substitution of a replicating artificial "RNA virus" for a more genetically stable DNA virus such as ASF (a large, double-stranded DNA virus in the Asfarviridae family) leads to more genetic instability. The cause of these deaths may not be the mishandling of the vaccine by farmers, as claimed.

Previous examples of problems with self-replicating RNA have been observed. Self-replicating translatable viral RNA has its dark side (the following data were collected by the Brooks AFRL Counterproliferation Team in 2000):

> The small and simple VEEV (Venezuelan equine encephalitis virus) has one more trick which it shares with the plant Tobacco mosaic virus; it can resurrect itself from broken pieces. The nsP4 sequence in VEEV is very similar to the replicases of three plant viruses, including tobacco mosaic virus. The Brooks Lab performed a critical experiment using the RetroXpress (TM) system of Clontech derived from a retrovirus that could transfer genes but could not infect and replicate. My group at Brooks constructed a new virus that contained the nsP4 sequence inserted into this platform and expected that the new virus could infect cells but not replicate. It not only infected (transfected) marker genes into the target mammalian cells but also formed plaques of transformed cells (those that resembled cancer cells and formed uninhibited growth mounds of cells). These cells grew and multiplied in the cell culture. A new completely infectious virus had formed. How? Tobacco mosaic virus

> has been shown to repair and reactivate after exposure to UV radiation with the help of an unknown plant nucleic acid repair system. Since we inserted only the nsP4 protein into the host cells coinfected with the defective retrovirus, we proposed that the new virus used cellular RNA-dependent RNA polymerase in conjunction with the expressed nsP4 to form a repair complex that constructed a chimeric virus from the retroviral pieces and other unknown viral nucleic acid components inherently present. Perhaps the infections by killed virus from killed vaccines previously reported were not just because of contamination with whole live virus but with components that survived the formalin damage and were reassembled into whole restored infectious virus, albeit at a less effective infectious dose of original broken pieces. Intact VEEV will produce a transmissible infection with as little as 12 PFU (plaque-forming units are equivalent to 10 virions for a multiplicity of infection of 1 but more likely 10X10 = 100 virions). I have speculated that these broken viruses, if reconstituted from broken virions or free RNA would require 10^6 units to make one intact virion—The Black Dragon Trilogy by JOHNATHAN KIEL https://a.co/gLQ6en9.

The Foundation for Food & Agriculture Research (FFAR) grant awarded to Genvax required collaboration with the USDA's Agricultural Research Services Plum Island Animal DiseaseCenter. In August 2022, before the Vietnamese pig deaths, Genvax received 6.5 million dollars from the coalition of meat packers, feed manufacturers, and other food supply chain groups to develop more self-amplifying vaccines. Other Genvax investors are the Iowa Corn Growers Association, Summit Agricultural Group, and Agricultural Startup Engine.

As mentioned earlier, we thought the delivery could be controlled electromagnetically because of the known mechanism of focusing E fields at the tips of high-aspect-ratio conductive fibers such as single-walled carbon nanotubes (SWCNTs) or multiwalled carbon nanotubes (MWCNTs) or DALM fibers or ribbons already observed, which facilitated the penetration of cell walls and membranes (a kind of wireless electroporation). MWCNTs have diameters between 7 and 100 nm, with aspect ratios between 50 and 4000. SWCNTs that are thinner (0.5–2.5 nm) have aspect ratios of up to 10,000. However, the "control penetration" of DALM/carbon nanotube/metallic nanoparticle composites (with no apparent applied nonionizing electromagnetic radiation (NEMR)) made that assurance moot. Other choices for control had to be pursued. One benefit of not using high-aspect-ratio CNTs is that we could avoid the inflammation associated with this CNT property and the potential for long-term pathology (resembling asbestosis or silicosis, to which they have been compared) [4]. However, unlike asbestos and silica, carbon nanotubes can be degraded by myeloperoxidase, limiting their long-term potential for pathology [4].

Another requirement was superimposed on these research and development goals. We wanted to use aptamers, as described earlier, not only for diagnostics but also in combination with DALM (perhaps also with CNTs and paramagnetic metallic nanoparticles) nanoparticle composites to neutralize pathogens and their toxins in the environment and, perhaps by being carried over, in human and animal hosts.

The ultimate goal was to develop pathogen and host-specific composite nanoparticles that could trace, identify, and neutralize pathogens and toxins seamlessly and autonomously. These notional composites were referred to as nanobes. These "little machines" controlled by RFR were comparable to the nanobots of science fiction. Following the requirements and needs of developing an aptamer against anthrax (actually the lethal toxin thereof) and other pathogens, we developed a formulation that could be delivered by self-administration through inhalation (Figs. 10.2 and 10.3) [5].

It would buy time for further treatment to extend the window of opportunity beyond the 4-day maximum before 100% death from biowarfare-inhaled anthrax spores. The preparations that preserved the aptamers from breakdown in the host and retained their effectiveness are described in a US Patent [5]. The results with Sterne strain and Ames strain anthrax in mice are graphically illustrated in Figs. 10.4 and 10.5.

One should note that these were in support of the antibiotic doxycycline, which by itself was insufficient for the 100% lethal doses of spores given by inhalation. The inhaled aptamers only neutralized the lethal toxin, not the bacteria. Therefore, continued bacterial growth in the mice, unless terminated by the mice's immune responses, would eventually overcome the mice, as observed. However, the initial goal was to self-deliver an inhaled countermeasure immediately after aerosol exposure to decrease casualties and improve the outcomes of subsequent medical treatment (buy time). The first step toward this goal was accomplished based on the data of overwhelming aerosol exposure of mice in two independent laboratories. The technology was an adaptation of a technology invented by Dr. Kary Mullis (Nobel Laureate for the invention of polymerase chain reaction (PCR); December 28, 1944–August 7, 2019; shared the 1993 Nobel Prize in Chemistry with Michael Smith and was also awarded the Japan Prize in 1993). The modified aptamers not only resisted nucleases but also redirected existing antibodies against agents, to

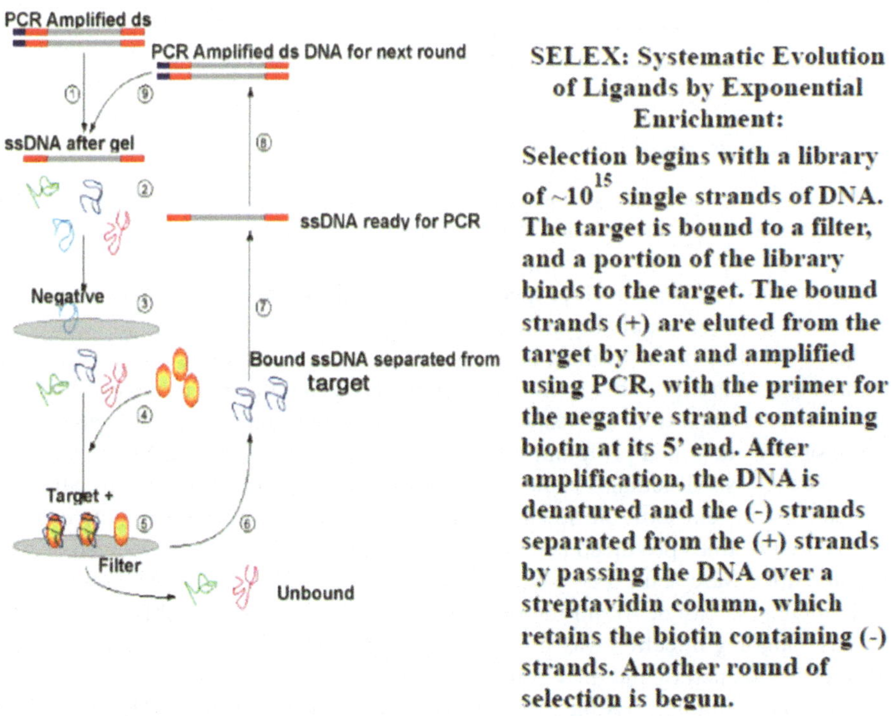

Fig. 10.2 The SELEX (systematic evolution of ligands by exponential enrichment) method for selecting aptamers

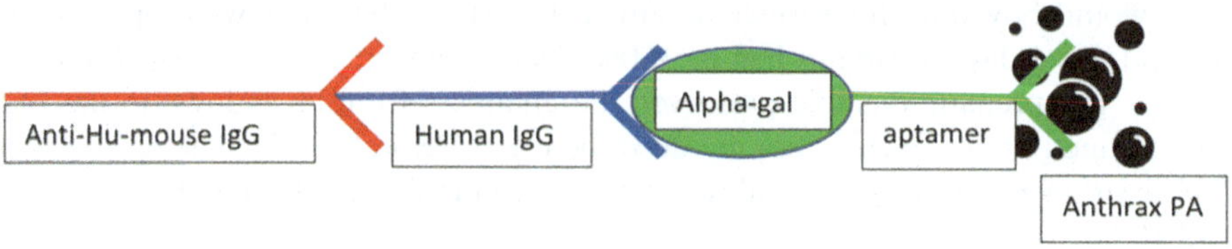

Fig. 10.3 Schematic of the antianthrax toxin chemically programmable aptamer composite for use in mice

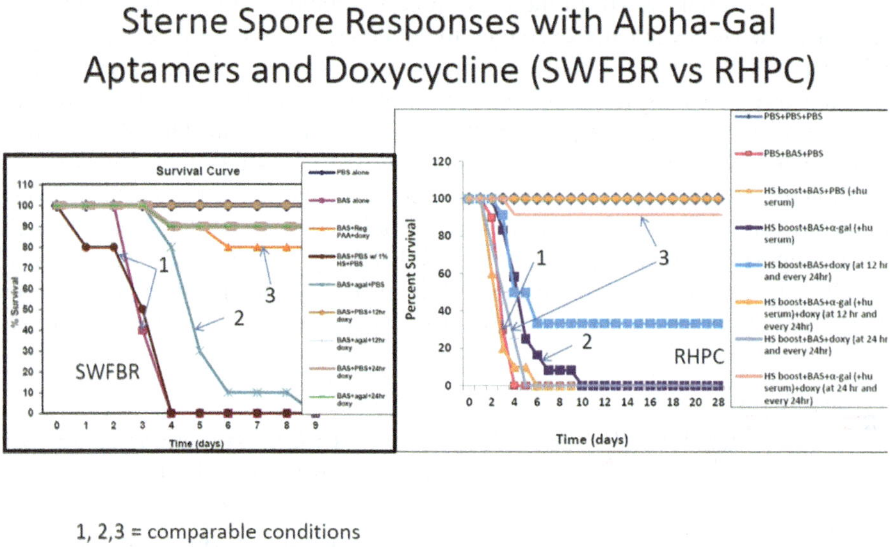

Fig. 10.4 Inhaled antilethal toxin aptamer effects on inhaled Sterne anthrax spore mortality in mice, with and without otherwise inadequate doses of doxycycline

which the host was immunized, specifically, toward new agents such as anthrax—a chemically programmable immunity.

From the data in the graphs, it can be noted that not all titrated lethal doses of anthrax spores are the same from lab to lab but are generalizable in response and in agreement in direction of response with the same treatments between labs. Additionally, there is a profound difference in the responses to the Sterne vaccine strain (which is still lethal for mice) and the very pathogenic Ames strain, which is difficult to prevent extreme lethal overdose. Even though the treatment of mice infected with Ames produced no cure, unlike infections with Sterne, delayed deaths in some mice were achieved. What is remarkable in the data is that when treatment with aptamers ceased, so did the delay to death. This result probably came about because ongoing infection was still yielding toxins, and the mice's immune systems were overwhelmed by the infection.

There is one other possible problem with aptamers that involves TLR (Toll-like receptor) 9 of the innate immune system. TLR9 can mediate either efficient Th1- or

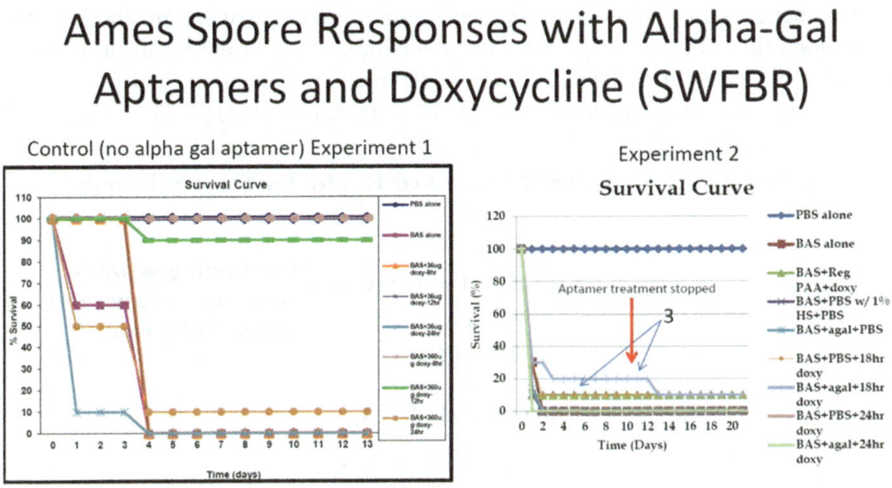

Fig. 10.5 Inhaled antilethal toxin aptamer effects on inhaled Ames anthrax spore mortality in mice, with and without inadequate doses of doxycycline

Fig. 10.6 TLR-9 non-CpG stimulators in antianthrax aptamers

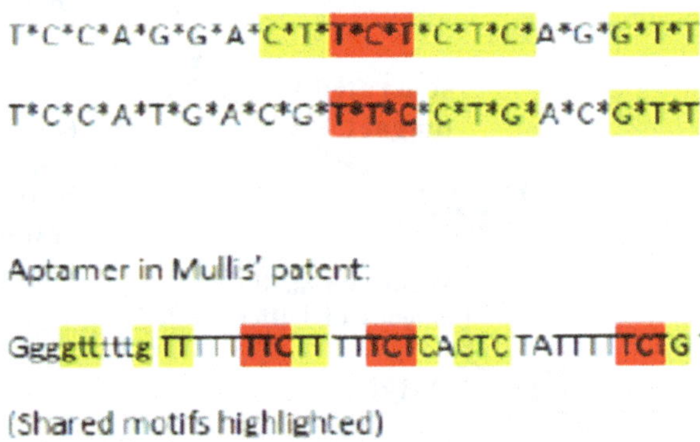

Th2-dominated lymphocyte immune effects, depending on whether it is stimulated by cytosine-phosphate-guanosine (CpG) or certain non-CpG oligodeoxynucleotides (ODNs). The non-CpG sequences that can be found in anti_anthrax toxin aptamers could cause this Th1 to Th2 shift, and the latter is usually considered less inflammatory. Figure 10.6 shows that some of the aptamers in the anthrax studies had these non-CpG sequences, TLR-9 non-CpG stimulators (phosphothiolated) [5].

The other problem was delivery. The aptamers were next incorporated onto the surface of nanoparticles by a covalent linkage so that they could be delivered by nanosprayers for the inhalation of particles of the size to enter the lungs.

The first control on their activity was chemical. The introduced nitrate reductase and the endogenous nitric oxide synthase form a hybrid redox chain that forms nitric oxide/nitrite from endogenous L-arginine and exogenous nitrate. The externally introduced synthetic DALM nanoparticle forms the new biosynthetic DALM in situ at the synthetic DALM nanoparticle's introduction and expression site.

Another control on direction and localization was the use of paramagnetic metallic nanoparticles in the composite. This magnetic control of the localization of the particles has other uses in the selection, replication, collection, and purification steps of the manufacturing process (Figs. 10.7, 10.8, and 10.9) [6, 7].

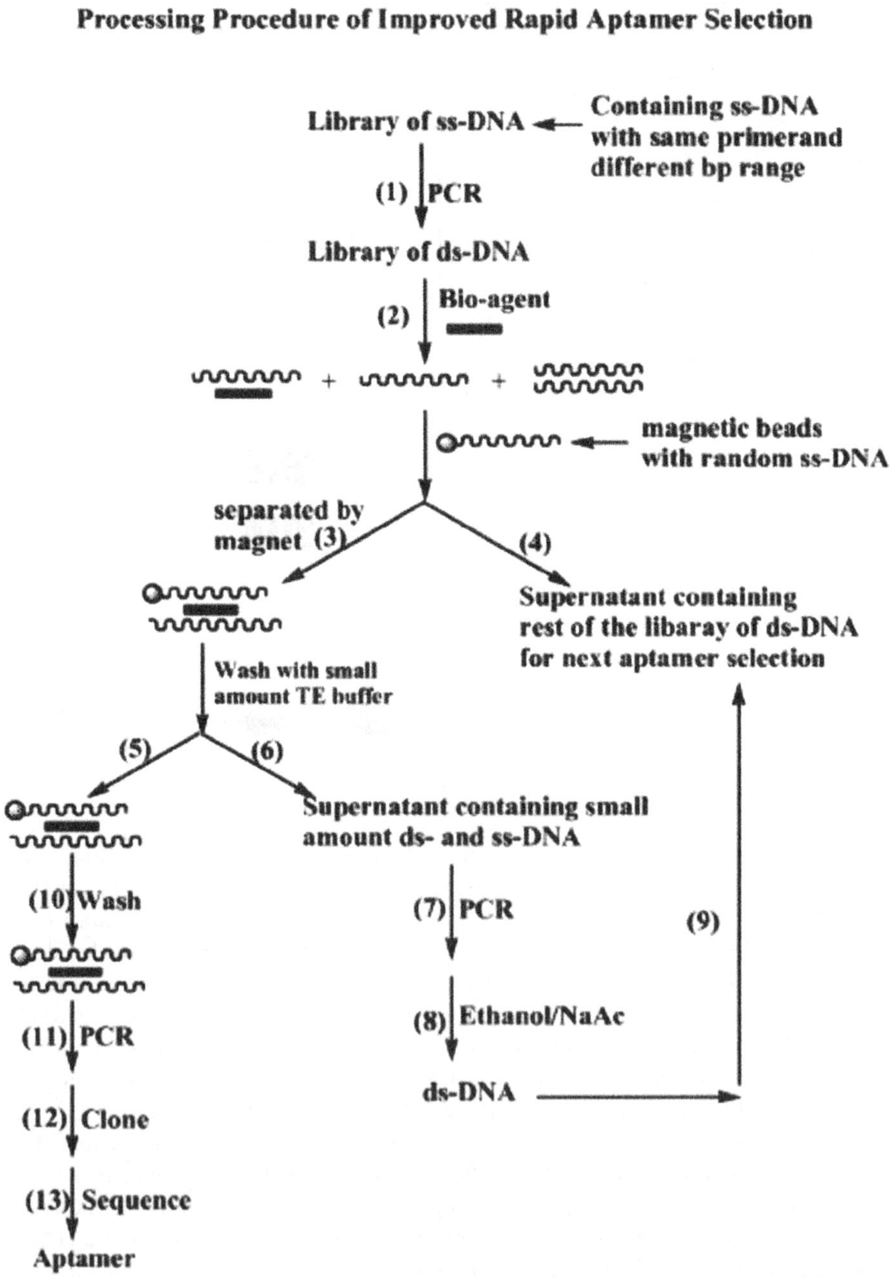

Fig. 10.7 Aptamer selection express (ASExp) method of selecting aptamers

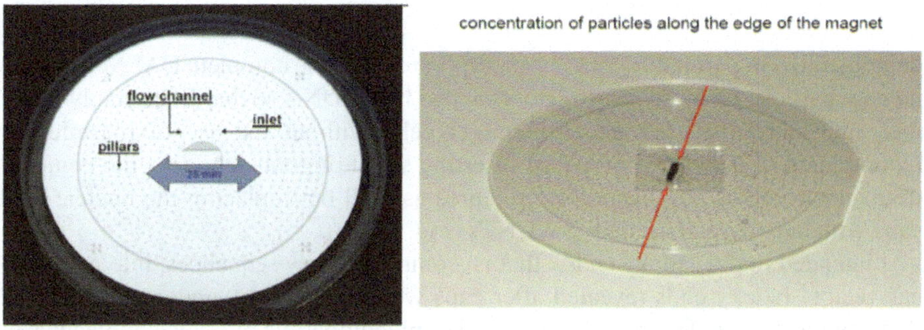

Fig. 10.8 First magnetic capture flow cell

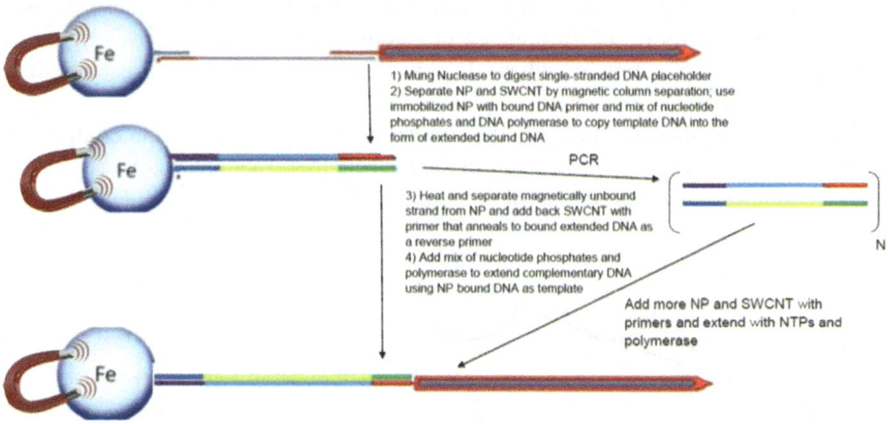

Fig. 10.9 Scale-up method for manufacturing synthetic nanoparticle/CNT aptamers

In addition to the manufacturing uses of magnetic fields and in spite of the lack of the need for microwaves or RFR to penetrate cells with the nanoparticle construct, NEMR is still necessary for disabling a virus or microbe and, if necessary, raising the level of radiation to the point of having a focused kill on the host cell. The induction of E fields and current by pulsed magnetic fields can also be used. It has the advantage of much less attenuation with depth of penetration than microwaves or RFR.

There are several proposed designs noted here and actually utilized composite nanoparticles.

Some, such as those previously demonstrated, have fluorescent tracers, dequenched when the nanoparticles are dissembled and activated. Those with magnetic nanoparticle components can be guided by an externally applied magnet field or activated by a pulsed magnetic field due to interaction of the metal with the triplet states and free radicals generated by the DALM. The metal, DALM, and CNTs also serve as contrast media markers of the intracellular locations of the nanoparticles in electron microscopy tissue sections, as noted earlier.

How can these composites become purely synthetic if they require plasmids as components to convey expressible genes? The answer may have already been demonstrated in part but is yet to be incorporated into the complete DALM composite nanoparticle. Chinese investigators capped linear DNA so that it can stably transfect and express marker genes in target cells without the use of plasmids or recombinant viruses [8]. I proposed inserting small oligonucleotides onto composite nanoparticle/carbon nanotube structures as well but replacing the nucleic acid caps with nanoparticles and/or SWCNTs (Fig. 10.10).

I intended to put "sticky ends," that is, complementary sequences, into microlinear vectors (sticky ends revealed after caps were removed by endonucleases) and then onto nanoparticles and SWCNTs. Other biocompatible biological molecules or specific aptamers would be placed on these structures as needed to facilitate cellular uptake through cell receptors or physical interaction with the specific cell wall components of the bacterial cells. Other components would be added to these surfaces to prevent the "crashing" of the nanostructures into agglomerates, conveying physical stability for application anywhere. Figure 10.11 is a simplistic representation of such a structure.

The design is also conducive to the in vitro mass replication of the composites. If successful in bringing all these components together, then completely synthetic DALM composite nanoparticles can be realized (Fig. 10.12).

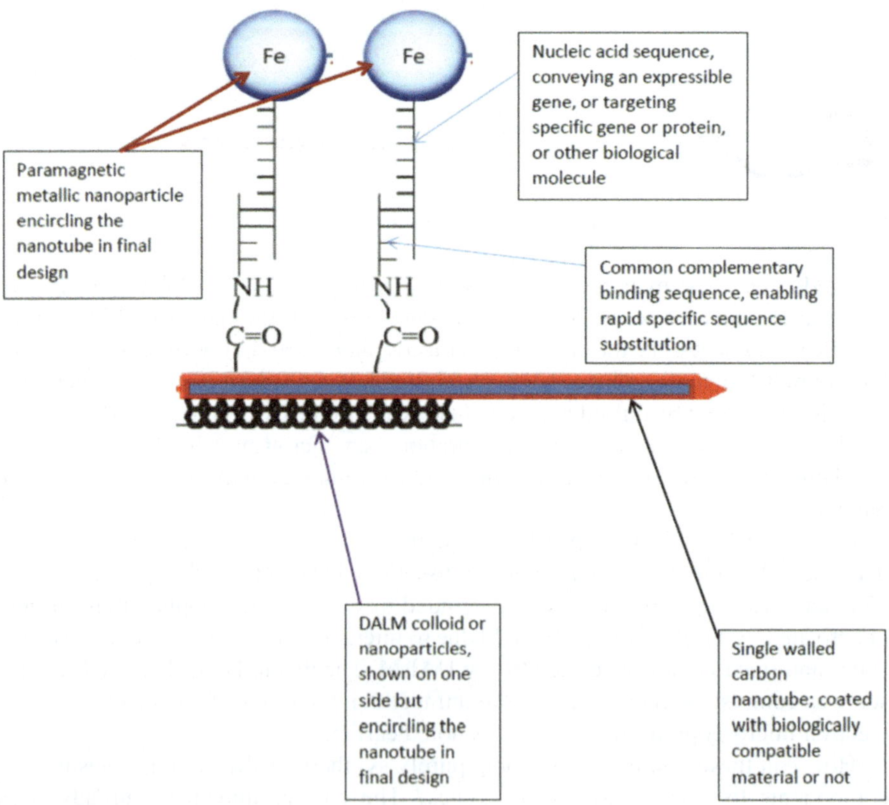

Fig. 10.11 Final synthetic DALM CNT/metallic nucleic acid nanoparticle design

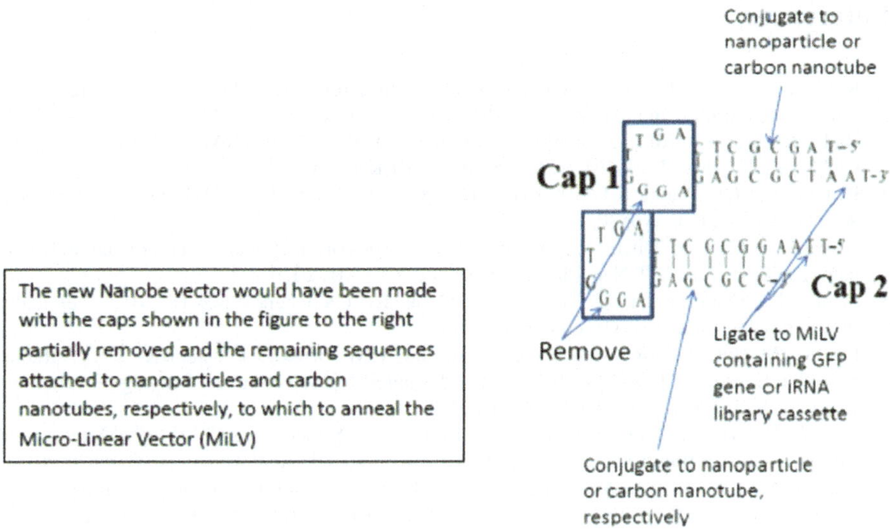

Fig. 10.10 Designing linear synthetic expressible nucleic acids for DALM synthetic CNT/ nanoparticle delivery

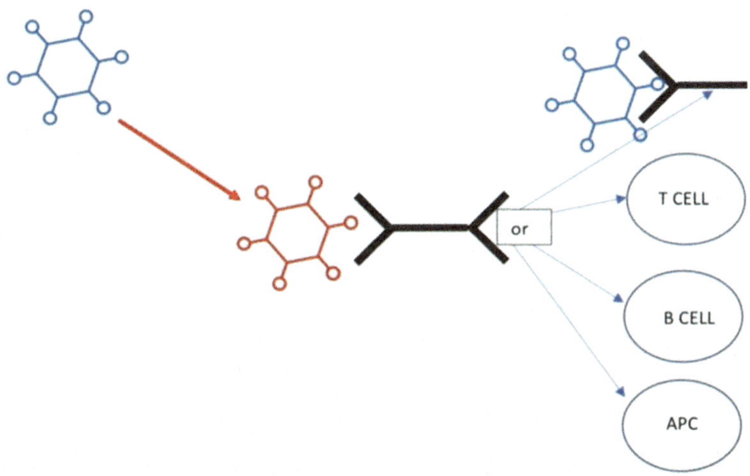

Fig. 10.12 How aptamers will redirect the humoral and cellular immune system to new viral variants after being delivered by the composite DALM nanoparticles. If the virus mutates into a new variant of concern (red), the original vaccine-induced immunity can be redirected by a bifunctional aptamer (selected in the test tube for binding to the new virus and the old viral antigen (blue, old virus)), with the old viral antigen attached to the aptamer and administered in a nasal spray; it can be self-administered upon a suspected exposure. The bifunctional aptamer with an old attached viral antigen links and redirects the old viral immune response (cellular and antibody) to the new virus (red). This approach allows us to outrun the evolution of the virus if the aptamers can be selected, manufactured, and distributed as close to the origin of the emergence of new variants quickly. The technologies I have described were designed for this very purpose [4]

References

1. Routledge, M., et al., Nitrite-induced mutations in a forward mutation assay: Influence of nitrite concentration and pH, Mutation Research/Genetic Toxicology 322(4): 341–346, 1994.
2. Radi, R., Peroxynitrite, a Stealthy Biological Oxidant, JOURNAL OF BIOLOGICAL CHEMISTRY 288(37): 26464–26472, SEPTEMBER 13, 2013.
3. Holwit, E., US Patent 6,379,900B1. COMPOSITIONS AND METHODS OF USE OF 8-NITROGUANINE, Apr.30, 2002.
4. Atiyah, A., Haide, A., Dhahi, R., Cytotoxicity properties of functionalised carbon nanotubes on pathogenic bacteria, IET Nanobiotechnol 13 (6): 597–601, 2019.
5. Mullis, K., Vivekananda, J., Kiel, J., and Cook, R., US Patent 8,604,184 B2, Chemically Programmable Immunity, December 10, 2013.
6. Fan, M., McBurnett, S. R., Andrews, C. J., Allman, A. M., Bruno, J. G., and Kiel, J. L.. Aptamer Selection Express: A Novel Method for Rapid Single-Step Selection and Sensing of Aptamers. J Biomol Tech 19(5), 311–319, December 2008.
7. Kiel, J., Holwitt, E., Fan, M. (Maomian), Roper, S., US Patent 9,273,345 B2. Methods and Compositions for Rapid Selection and Production of Nucleic Acid Aptamers, March 1, 2016.
8. Wang H-S, et al., A Novel Micro-Linear Vector for In Vitro and In Vivo Gene Delivery and Its Application for EBV Positive Tumors. PLoS ONE 7(10): e47159. doi: https://doi.org/10.1371/journal.pone.0047159, 2012.

Chapter 11
Future Applications Beyond Microbial Killing: Promise or Peril?

Abstract This chapter is mostly for biomedical engineers and biotechnologists. It is about the future, which is risky to predict at best, an ever-moving target. The technology of radio-controlled autonomous vectors has had a rough start. It has been hidden in government laboratories and published and patented in obscurity. It has been met with suspicion and distrust based on the field out of which it emerged and its dual and nefarious potential use. However, like all disruptive, revolutionary technologies, it will be under suspicion until it is in wide use for at least some common applications. Meanwhile, I am passing the baton to those brave and clever enough to pursue it. We will now let our imaginations take us to future applications.

Keywords Microwave hearing · Thermoacoustics · Transcranial magnetic stimulation · Crops · Pesticides · Transfection · Etching · Plastic · Bubbles

Discovery can be an exhilarating experience or a terrible and frightening thing. When I was in Madrid, Spain, in 2004, I went to the Plaza Mayor to buy my son a sword, a rapier, made by a famous manufacturer in Toledo, Spain. Walking across the Plaza Mayor, the history was palpable. It was as if I could still hear the clicking of the Spanish spurs of the Conquistadors on the pavement. The plaza dates back to the fifteenth century, when it was originally called the "Plaza del Arrabal." In 1561, the plaza came under the jurisdiction of the city of Madrid. King Philip II of Spain commissioned Juan de Herrera to redesign the plaza. This was during the pinnacle of Spain's conquest of the New World and the growth of the Empire. King Philip III began the reconstruction in 1617, and it was completed in 1619. Spanish discovery and greatness brought disease, death, suffering, misery, and slavery to Meso America (Olmec, Zapotec, Maya, Toltec, and Aztec peoples) and other native populations (the Inca of Peru) and slavery from Africa to the New World. The beautiful pre-Colombian edifices were torn down and built over or abandoned in the jungle. Engineering was not as much a part of this conquest as it was for the Romans, where military conquest was intimately tied to engineering and the "Pax Romana." Nonetheless, even the success of science is tied to the implementation by

J. L. Kiel, *Bionanotechnology*, https://doi.org/10.1007/978-3-031-63006-4_11

engineering and its technological and financial success, which must come with caveats and perils as well as an exhilarating promise.

This chapter is more focused on engineering and practical solutions with caveats. Engineering is about designing machines and structures, the artful application of science and mathematics through properties of matter and sources of energy to make objects and processes that have practical applications for a wide range of people. Although some may construe that this chapter is only speculative, they would be in error because all the components are well grounded in scientific data and theory, which I have demonstrated in the previous chapters. The combinations into practical applications, limited by financial support and the will to make them, are the only speculations.

The caveats are rooted in the worst-case scenarios growing out of basic science possibilities. Therefore, we will address some of these first before proceeding to practical applications. The neurological, beyond what I have already said about basic metabolic and genetic effects on the molecular level, is what we will focus on here.

Nitric oxide (NO) binds to the heme group of proteins, such as guanyl cyclase and hemoglobin. Most neurological effects of NO are due to the activation of guanyl cyclase, in turn producing guanosine 3′,5′-cyclic monophosphate (cGMP), but there are also other guanyl cyclase-independent mechanisms at play in the systemic and brain functions [1, 2].

Different types of nitric oxide synthase (NOS) exist from that in endothelial cells to inducible NOS of inflammatory cells to that in astrocytes, microglia and vascular cells in the brain, and four types of neuronal NOS. This enzyme is found in the cerebellum and many other parts of the brain, where it functions in neuronal differentiation, pain, neuroendocrine function in the hypothalamus-hypophysis axis, sleep and hippocampal function. The four types of neuronal NOS are nNOS alpha and nNOS micron, intracellular membrane associated, and nNOS gamma and nNOS beta, cytoplasmic. nNOS alpha is attached to the postsynaptic nerve ending by postsynaptic density protein 95 (PSD-95), which mediates interaction with N-methyl-D-aspartate (NMDA) receptors. The N-methyl-D-aspartate receptor is a glutamate receptor. L-glutamate receptors play an important role in learning, memory, and, in particular, spatial memory.

Thus, the diazoluminomelanin (DALM) composite nanoparticles could control these neurological functions by placing the nanoparticles in the preferred location and triggering their release of NO by providing precursors such as nitrate. However, there is an alternative that directly engages the physics of nonionizing electromagnetic radiation (NEMR).

The current wireless technology for stimulating the central nervous system is repetitive transcranial magnetic stimulation (rTMS). Noninvasive TMS uses pulsed magnetic fields to stimulate neurons in the brain to improve symptoms of depression. It is unlike electroconvulsive therapy (ECT), which causes seizures and requires sedation and anesthesia. However, it can cause side effects, including headaches, scalp irritation; tingling, twitching, or spasms of facial muscles; lightheadedness; and, on rare occasions, seizures, mania, especially in bipolar patients, and hearing loss if the ears are not protected during treatments. Such a device could be used to activate DALM composite nanoparticles by induction after they reach their target.

A magnetic wireless noninvasive diagnostic technique, that could be used to monitor DALM magnetic nanoparticles in the brain, is implemented by a superconducting quantum interference device (SQUID), an ultra-sensitive magnetometer [3]. The device is based on two superconductors separated by narrow insulating layers that compose two parallel Josephson junctions. These junctions are subjected to a constant biasing current so that voltage oscillations can be measured with changes in phase at the two junctions, which depends upon the change in magnetic flux. Counting oscillations facilitates the evaluation of flux changes. Flux is quantized in quantum units. Equation 11.1 describes this relationship for a single magnetic flux quantum.

Equation 11.1: Single Magnetic Flux Quantum

$$\Phi_0 = \frac{2\pi\hbar}{2e} \cong 2.0678 \times 10^{-15} \text{ tesla} \cdot \text{m}^2$$

The two superconductors in the SQUID separated by a narrow insulating layer experience a tunneling of Cooper pairs of electrons (pairs of electrons with opposite spins and moving at the same speed in opposite directions, to which the phenomenon of superconductivity is attributed) through the junction. This effect is the basis for the sensitivity of the device. The SQUID sensitivity has a threshold of 10^{-14} Tesla. This is in a range comparable to the magnetic field of the heart, 10^{-10} T, and of the brain, 10^{-13} T. The Cooper pairs on each side of the junction are represented by wave functions similar to free-particle wave functions. The direct current (DC) Josephson effect is a current proportional to the phase difference of the wave functions flowing through the junction in the absence of voltage. The alternating current (AC) Josephson effect is Josephson junction oscillation at a characteristic frequency that is proportional to the voltage across the junction. This device could probably measure the induced magnetic fields of the radiofrequency radiation (RFR)-activated DALM composite nanoparticles to assess their level of activity and location in the body, even the brain.

Delivery to the brain does not necessarily require intravenous or intracranial injection. There is evidence that nanoparticles can reach the brain through the nose. Wang et al. [4], demonstrated entry into the central nervous system by instilling TiO_2 nanoparticles intranasally in mice [4]. They observed changes in the cells of the olfactory bulb and hippocampus regions of the brains of mice exposed to the nanoparticles but not in the cerebral cortex or cerebellum. The number of cells in the olfactory bulbs actually increased, while the cells in the hippocampus were damaged and degenerated. They and others also observed increased inflammation and cell stress in exposed mice [4, 5]. The composite DALM/carbon nanotube (CNT)/paramagnetic nanoparticles, as conductors, could enter the brain by inhalation through the nose and lodge in areas in the brain where the application of pulsed magnetic fields could induce a current that would alter the activity of nerves in highly localized regions. By adding specific aptamer sequences to the surface of the particles, they could be directed to bind to specific neurons to deliver "nanolevel stimulation." This application would have dual use, with the nefarious one being the behavior control of specific adversaries by the external application of electromagnetic fields that would not affect anyone else exposed. However, this approach could also provide relief for those suffering from various neurological or mental diseases and brain injuries, including counteracting addiction. However, this approach presents a dual-use dilemma.

We attempted to put the potential central nervous system (CNS) applications to good use by answering a solicitation from DARPA (Defense Advanced Research Projects Agency; November 12, 2013). The ultimate intention of the proposed research was to provide minimally perturbing implants of electrical stimulators and recorders of neurological activity in the central nervous system to meet the DARPA Broad Agency Announcement (BAA) goal to deliver a platform technology for precise therapy in humans living with neuropsychiatric and neurologic disease, including veterans and active-duty soldiers suffering from mental health issues associated with war injuries. The proposal intended to generate methods that would be used in neural recording and stimulation to close the loop on therapeutic treatment in individuals who receive minimal benefits from currently available treatments. The research and development were intended to improve knowledge of multiple neural subnetworks of the brain that are involved in disease and illness by providing nanoscale precision in stimulating and recording neurons. This proposal combined novel device development with state-of-the-art nanoscale bioelectromagnetics. Because of the unprecedented novelty of wireless nanoelectrode device(s), we did not anticipate the immediate use of the technology in the CNS until extensive preclinical and clinical investigations were completed. However, therapeutic cranial nerve stimulation to treat paralysis or, at higher energies, nerve ablation to treat hyperkinesia and the treatment of diabetic neuropathy by local electrostimulation with nanoelectrodes are considered near-term commercial biomedical applications. These wireless nanoelectrodes could be energized by external devices that are not limited by size or energy sources, as implantable systems have been. Unfortunately, the response was negative, and another good idea for the application of DALM composite nanoparticles went into the dustbin.

The next application to consider is "genetic surgery," cutting out and replacing genes. DALM nanoparticles may have already demonstrated this capability, based on the previous data presented, when they removed bacteriophage susceptibility and the suppression of hemolytic activity in *Bacillus anthracis*. However, directed cutting of certain sequences without wholesale destruction of the target cell requires further refinement. A two-nanoparticle tandem approach to applying clustered regularly interspaced short palindromic repeats (CRISPR) for this purpose may be possible, as illustrated in Fig. 11.1.

Figure 11.2 illustrates the sophisticated approach to removing toxin genes and antibiotic resistance genes from a targeted animal pathogen, *Pasteurella multocida*,

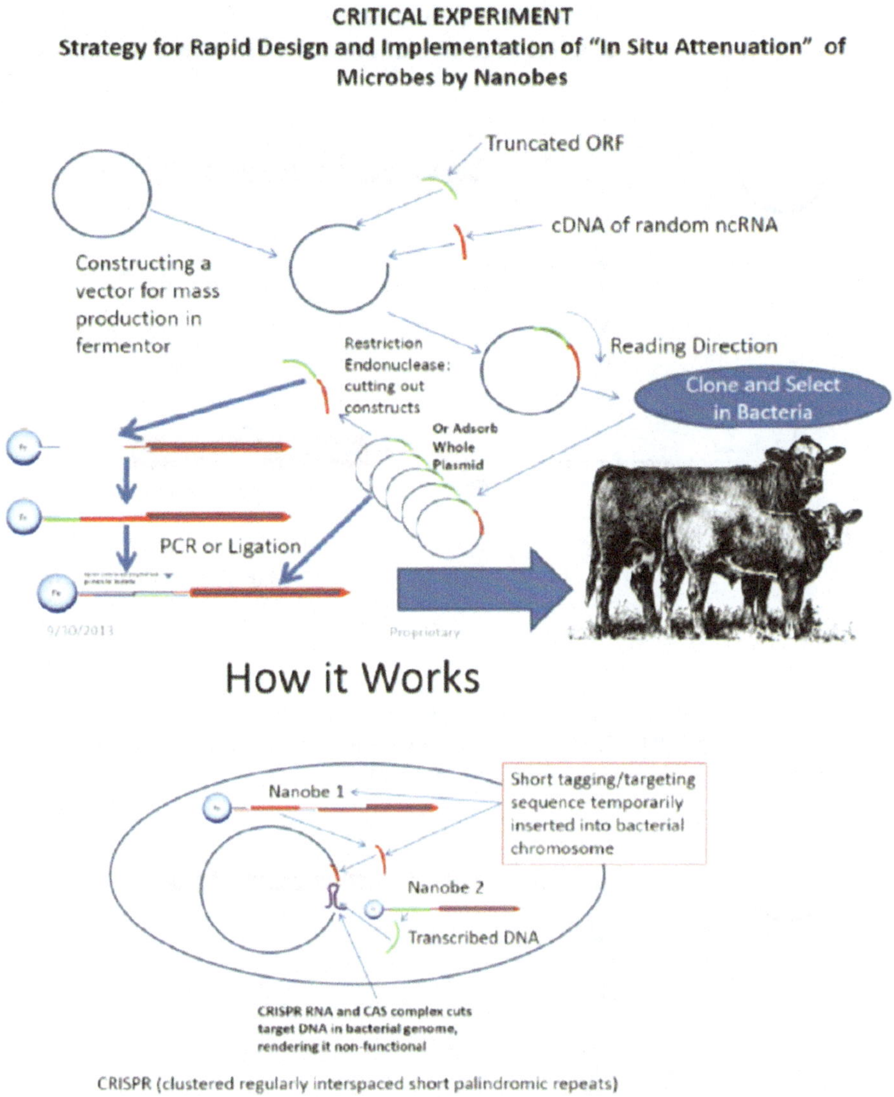

Fig. 11.1 Fundamental design of binary aptamer synthetic nanoparticles for the delivery of CRISPR

which may be executed extracorporally (in the environment) or in vivo in an infected animal.

However, this approach to fulfilling the present need for a highly targeted delivery of CRISPR components may be far more complicated than is probably needed.

DALM, as I have demonstrated, can bind to deoxyribonucleic acid (DNA) and could, in turn, use this DNA to specifically direct binding to complementary sequences in a chromosome or ribonucleic acid (RNA) in the nucleus or cytoplasm

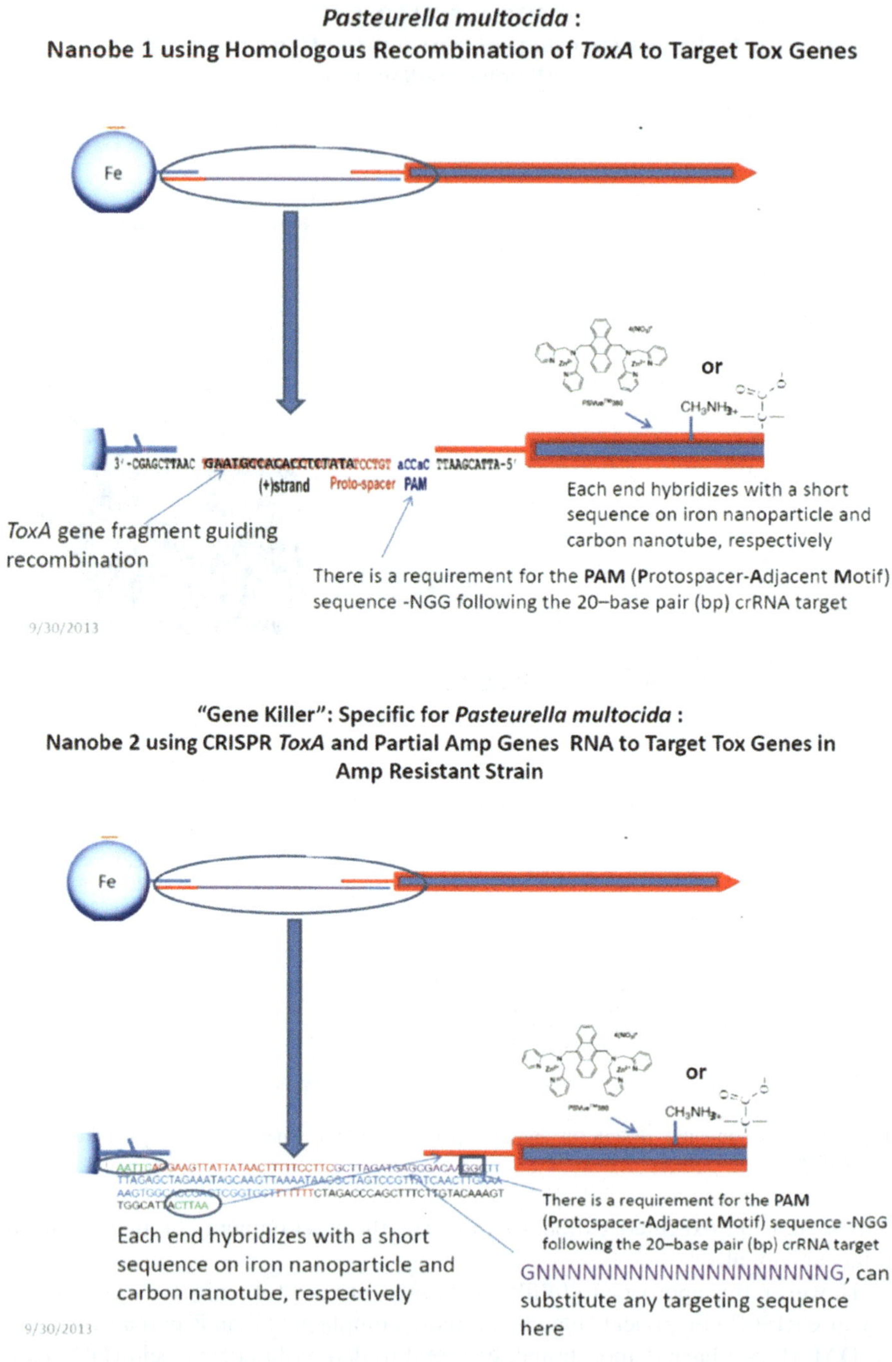

Fig. 11.2 Specific design of binary aptamer nanoparticles for the CRISPR treatment of *Pasteurella multocida* to remove toxin and antibiotic resistance genes associated with antibiotic-resistant shipping fever in cattle

of a cell, respectively. When the redox function of DALM is activated by the application of light and/or electromagnetic radiation, the activity cuts or damages the DNA significantly at that location, leading to its removal by DNA repair enzymes. Other DALM DNA combinations, with DNA complementary to flanking sequences, could fill in the damaged area in a host by a homologous recombination repair process, effectively cutting and repairing the damaged DNA with a new sequence and generating a new expressible trait. Therefore, the DALM composite nanoparticles could already make CRISPR obsolete, not requiring the two-component approach of CRISPR. This CRISPR-like activity could change heritable traits and open a wide range of gene therapies and genetic modifications controlled by RFR, pulsed magnetic fields, or both.

There are many potential nonmedical uses of DALM composite nanoparticles: (1) transformation vector for agriculture, (2) RFR (and other NEMR)-controlled pesticides, (3) transient persistent transformation of crops (for one season; self-limiting desirable trait transfer), (4) transformation vector for bioenergy, (5) enhancing cyanobacterial production of biofuels, and (6) enhancing microbial production of materials and feedstocks. Furthermore, when activated with high-peak-power pulsed microwaves, the nanoparticles form highly localized high temperature but very brief nanoplasmas. This highly focused application produces a high-resolution etching of materials. Initially, our investigations of these characteristics led to suspicion that military antimaterial use was being pursued in violation of international law and treaty, which includes antimaterial bioagents, along with typically especially dangerous pathogens (Fig. 11.3).

Fig. 11.3 Effects of repeatedly or continuously activated DALM (as by light) on materials such as epoxy and fine oils

However, civilian utility should not be abandoned: (1) nanoplasmic cleaning of surfaces through E-field focusing, (2) nanoplasmic etching of surfaces for nanodevice construction, and (3) the breakdown of environmental plastics.

Finally, we must consider the production of sound by DALM nanoparticles. The previous discussions were of CNT fibers that vibrated as taunt "strings" between electrodes when exposed to RFR signals, nanoradio. The biological mechanism of producing sound with microwaves is microwave hearing. It is the result of the thermoacoustic effects of pulsed RFR on the tissues of the skull interacting with the inner ear and the thermoelastic expansion of the components of the auditory sensing organ. This mechanism has moved to the forefront of clandestine, unexplained counterintelligence international incidents. Havana syndrome (first observed in 2016 among US diplomats in Havana, Cuba) is manifested by a range of symptoms, including dizziness, headache, fatigue, nausea, anxiety, cognitive difficulties, and memory loss of varying severity. In some cases, diplomats and intelligence officers have left active service due to complications from the condition. Was it caused by pulsed microwaves? The Central Intelligence Agency (CIA) reported that there was no evidence linking Havana syndrome to a foreign country (January 20, 2022, 9:43 AM ET NPR. Greg Myre National Security Correspondent). Truly? That does not say it was not done or done by a foreign power. It means they do not know or would not tell.

Foster et al. [6] posed the question of whether the auditory effect can be "weaponized," i.e., used to harass or harm an individual. They stated that due to the open scientific literature state of knowledge, using pulsed microwaves was unlikely, but the lack of publicly available information about existing high-power RF technology and uncertainties about thresholds for adverse effects did not allow for definitive conclusions. The question remained: Can the microwave auditory effect be "weaponized?" [6, 7].

First the delivery of DALM nanoparticles to targeted individuals as aerosolized nanoparticles clandestinely and then generalized exposure to RF sources would lower the threshold of bioeffects on the targeted individuals and spare the rest (unnoticed) who were RFR exposed. This enhancement would greatly reduce the required size of the transmitter and the limitations on the frequencies because of the depth of penetration and power requirements. Once triggered, these nanoparticles may trigger residual effects without additional RFR exposure, meeting Dr. Foster et al.'s requirements for "weaponization." The DALM metallic nanoparticles/pulsed microwave potential weapon or a similar structure will meet the criteria if an alternative mechanism exists.

Neither of the mechanisms (thermoacoustic nor nanoradio) explains DALM nanoparticle sound generation from pulsed microwaves [8]. Thermacoustics involves relatively small temperature and pressure changes over small penetration depths, on the order of 0.1 mm (100-micron range), mostly at gas/solid interfaces. Thermoacoustics devices can generate a sound level of up to 180 dB (whispering at 30 dB, conversation at 60 dB, above 70 dB—prolonged exposure damages hearing—and above 120 dB—immediate damage to hearing), with pressure variations of 30 kPa (0.296 atmospheres). The idea that DALM fibers vibrate like musical strings did not hold up once the interacting structures were found to be spherical nanoparticles. The high-temperature electronic light emissions were more in accord with vibrating bubbles, as seen with sonoluminescence. The temperatures and pressures in the bubbles are extremely high and not suppressed by the heat capacity of water vapor, as evidenced by the sharp light emission spectra observed rather than broadened light emission spectra. This isolation of the hot gas in the bubble from the aqueous solution, oddly enough, is due primarily to thermal conduction, which allows for the further compression and heating of the bubble without having to fight the energy-sapping heat capacity of water (mass transfer of water vapor into the bubble) [9, 10]. The high-temperature effects manifested in the killing of bacterial spores (the "popcorn effect") confirmed this heat transfer hypothesis. The sound is not just the production of "miniature thunder" with the plasma electrical discharge ("lightning in a bottle") but also the vibrations of the plasma gas in the bubble electromagnetically coupled to the microwave pulses [11, 12].

Thus, nanoplasmas also play a key role in sound production. This might be called the "hot nano lightning in a bubble" mechanism.

In review, the paradox is how can a microwave effect that mimics a thermal effect work best only when the object being irradiated is cooled? The effect is the killing of heat-resistant spores in very short time frames that mimic killing at very high temperatures at short time exposures. The luminescence of activated DALM that correlates with microwave or RFR heating does not correlate with microbial killing. However, the flashes and sound produced with very high peak power correlate with very-high-temperature plasmas that kill these spores instantaneously. The molecular mechanism of killing is dependent on free radical generation from an electromagnetic-field-facilitated conversion of chemically or photochemically formed singlet states to triplet states, which then yield free radicals. Therefore, the molecular mechanism of this type mimics ionizing radiation effects rather than the classical thermal mechanisms of RFR and microwave bioeffects [13].

The sound is produced not by the resonances of taut fibrous structures such as CNTs but by spherical nanoparticles surrounded by vibrating plasma bubbles. Although I stated previously that the sound was not exclusively from the electrical discharge above the DALM solution with microwave pulses, inside the bubbles, lightning-like discharges can occur and, as we have seen, produce very high temperatures in very small volumes for very brief periods of time. Similar to the thunder of lightning, the electron temperature of the gas inside the bubbles and the discharge channel within may reach as high as 50,000 °F, five times hotter than the surface of the sun, rapidly expanding the bubble. After the flash, the gas would cool and contract suddenly. The light pulses (see Fig. 3.5) precede the sound pulses, similar to lightning preceding thunder, supporting this mechanism [11]. This rapid expansion and contraction would create a sound wave that resembles thunder "in a bottle" [12].

This is the question: Is this the beginning or the end of radio-controlled bionanotechnology? The search for a "film badge" approach has not ended even though DALM's has [14]. The pursuit of magnetically guided nanoparticles in tissues to enhance genetic therapeutic specificity continues [15]. Although there are some custom specialty chemical sources of DALM, the only commercial source, if you

Fig. 11.4 Three forms of DALM polymer: (1) original DALM (ribbon and triclinic crystalline form), (2) paramagnetic iron micro–/nanoparticle DALM form, and (3) DALM-coated polystyrene. As the polymer "matures," it turns pure black. (Courtesy of the J&R Project)

care to investigate the properties and applications of this unique polymer, is the J&R Project (Fig. 11.4). If one wants to make the original material, then it can be prepared by combining solutions of 10 mM luminol in dimethyl sulfoxide (DMSO), 10 mM 3AT in water, and 100 mM $NaNO_2$ in water. The orange-brown solution contains DALM, which is quantified by its absorbance at 500 nm wavelength light. The water-soluble polymer (colloidal) is then precipitated with acetone and air-dried. This seems ridiculously simple for 30 years of research, but its interaction with nonionizing electromagnetic radiation has proven to be far more complex than just finding the "magic" frequency of microwaves or radiofrequency radiation to create a nonthermal bioeffect.

If there were any doubt that the technology described in this chapter is not speculation, then the final experiments to prove the killing power seen in the military research, including those reproduced in a low-cost private laboratory, shown in Fig. 11.5 should dispel it. Figure 11.5 shows data from the pictured laboratory water flow system illustrating the killing of the surrogate bioagent *Bacillus thuringiensis* (spores) in 24 h using ambient energy sources without utilizing expensive microwave emitters.

Figure 11.6 shows a fluorescent photomicrograph of a totally synthetic nanoparticle transfection of targeted mammalian cells to express green fluorescent protein

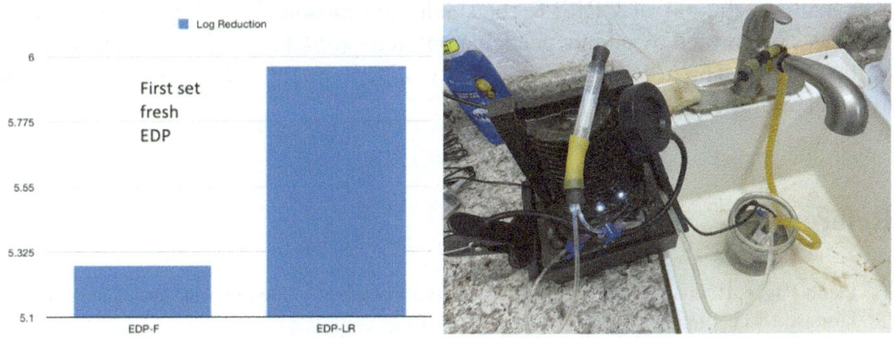

Fig. 11.5 Log reduction in recoverable *Bacillus thuringiensis* from the fountain (F) and large reservoir (LR) in the flow system pictured after 24 h of passive treatment with coupons of epoxy/DALM/iron nanoparticles (EDP) on transparent polystyrene. (Courtesy of the J&R Project)

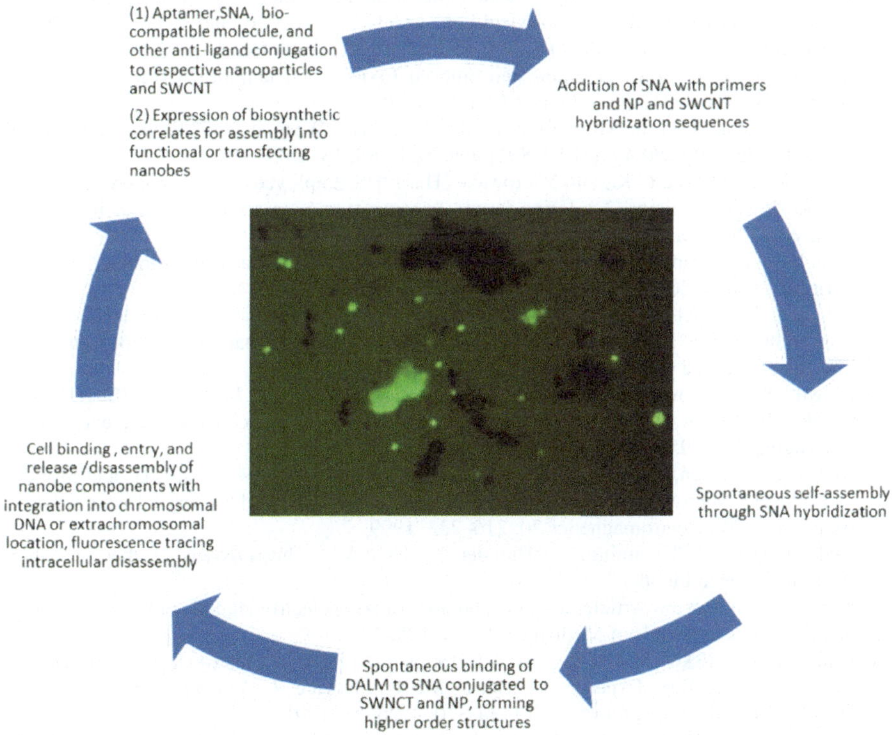

Fig. 11.6 The cycle of totally synthetic nanobes to totally self-replicating biosynthetic nanobes. *SNA* small synthetic nucleic acid, *SWCNT* single-walled carbon nanotube, *NP* nanoparticle, *DALM* diazoluminomelanin; the photomicrograph is of live cells in culture expressing green fluorescent protein (GFP) after transfection with totally synthetic nanobes carrying the gene

even without the aid of CRISPR for stable chromosomal integration. This latter work was performed by an independent contractor (apart from the author) on behalf of the Defense Threat Reduction Agency, US Department of Defense, a final proof of concept of the bionanotechnology.

References

1. Picón-Pagès, P., et al., Functions and dysfunctions of nitric oxide in brain, Molecular Basis of Disease 1865: 1949–1967, 2019.
2. Grover, S., Long-lasting, dissociable improvements in working memory and long-term memory in older adults with repetitive neuromodulation, Nature Neuroscience, https://doi.org/10.1038/s41593-022-01132-3, www.nature.com/natureneuroscience, 2022.
3. Körber, R., et al., SQUIDs in biomagnetism: a roadmap towards improved healthcare, Supercond. Sci. Technol. 29: 113001 (30pp), 2016.
4. Wang J, et al., Potential neurological lesion after nasal instillation of TiO_2 nanoparticles in the anatase and rutile crystal phases. Toxicol Lett. 183(1–3):72–80. doi: https://doi.org/10.1016/j.toxlet.2008.10.001. Epub 2008 Oct 17. PMID: 18992307, 2008 Dec 15.
5. Jia et al. The Potential Liver, Brain, and Embryo Toxicity of Titanium Dioxide Nanoparticles on Mice, Nanoscale Research Letters 12:478, 2017.
6. Foster, K., et al., Can the Microwave Auditoy Effect Be "Weaponized"? Front. Public Health 9:788613. doi: https://doi.org/10.3389/fpubh.2021.788613, 2021.
7. Foster, K., and Chou, C-K., Did Microwaves Harm US Employees at Its Embassy in Havana? Science-Based Medicine: Exploring issues & controversies in science & medicine, https://sciencebasedmedicine.org/, May 26, 2021.
8. Tian, H., et al., Graphene Earphones: Entertainment for Both Humans and Animals, ACSNano 8(6): 5883–5890, 2014.
9. Lebedev, Y., Microwave discharges: generation and diagnostics, Journal of Physics: Conference Series 257: 012016, 25th Summer School and International Symposium on the Physics of Ionized Gases—SPIG 2010 IOP Publishing, 2010.
10. Brown, S., Microwave Studies of Gas Discharge Plasmas, MICROWAVE STUDIES O F PLASMA, 394-397, http://www-naweb.iaea.org/napc/physics/2ndgenconf/data/Proceedings%201958/papers%20Vol32/Paper49_Vol32.pdf.
11. Kiel, J. L., Seaman, R. L., Mathur, S. P., Parker, J. E., Wright, J. R., Alls, J. L., and Morales, P. J. Pulsed Microwave Induced Light, Sound, and Electrical Discharge Enhanced by a Biopolymer. Bioelectromagnetics 20: 216–223, 1999.
12. Understanding Lightning: Thunder, NOAA, https://www.weather.gov/safety/lightning-science-thunder.
13. Sowa, P., et al. Review Article: Ionizing and non-ionizing electromagnetic radiation in modern medicine, Polish Annals of Medicine 19: 134–138, 2012.
14. Dmitriev, A.S., Itskov, V.V. & Ryzhov, A.I. Personal Dosimetry of Microwave Electromagnetic Radiation. Long-Term Experiments with a "MERA" Device. *J. Commun. Technol. Electron.* 66, 23–33 (2021). https://doi.org/10.1134/S1064226921010010.
15. Trueck, C., Zimmermann, K., Mykhaylyk, O. et al. Optimization of Magnetic Nanoparticle-Assisted Lentiviral Gene Transfer. Pharm Res 29, 1255–1269 (2012). https://doi.org/10.1007/s11095-011-0660-x.

Correction to: Nonionizing Electromagnetic Radiation-Driven Microbiological Effects

Correction to:
Chapter 8 in: J. L. Kiel, *Bionanotechnology*, https://doi.org/10.1007/978-3-031-63006-4_8

The original version of this chapter "Nonionizing Electromagnetic Radiation-Driven Microbiological Effects" was inadvertently published with incorrect heading in column 5 of Table 8.2. The column heading has now been corrected in the chapter.

The updated version of this chapter can be found at
https://doi.org/10.1007/978-3-031-63006-4_8

J. L. Kiel, *Bionanotechnology*, https://doi.org/10.1007/978-3-031-63006-4_12

Index

J. L. Kiel, *Bionanotechnology*, https://doi.org/10.1007/978-3-031-63006-4

The manufacturer's authorised representative in the EU is Springer Nature Customer Service Centre GmbH, Europaplatz 3, 69115 Heidelberg, Germany. If you have any concerns regarding our products, please contact ProductSafety@springernature.com

Printed and bound by CPI Group (UK) Ltd, Croydon, CR0 4YY
28/09/2026
02224081-0004